Synthese Library

Studies in Epistemology, Logic, Methodology, and Philosophy of Science

Volume 521

Dena Shottenkirk

Art as Cognition: How Gist Reframes the Aesthetic Experience as Conversation

Springer

Dena Shottenkirk
Brooklyn College (City University of New York)
Brooklyn, NY, USA

ISSN 0166-6991 ISSN 2542-8292 (electronic)
Synthese Library
ISBN 978-3-032-00142-9 ISBN 978-3-032-00143-6 (eBook)
https://doi.org/10.1007/978-3-032-00143-6

This Springer imprint is published by the registered company Springer Nature Switzerland AG
The registered company address is: Gewerbestrasse 11, 6330 Cham, Switzerland

To Frankie: my conversation and drawing partner

Preface

This book was not my idea. Although that is probably true for a lot of books (as ideas sometimes tend to be handed back and forth and the seed for a big project like a book might get planted in someone's mind almost without them knowing it), I will have to be honest and say that in my case I know quite well where this idea originated as I still have that email. The idea came from an editor who, while rejecting the proposal I had sent, offered some advice instead. She wrote, "As a successful, active artist and a philosopher, I imagine you have a very unique understanding and way of thinking about art and philosophy. Would a project about, say, the way philosophy informs your art be in the realm of possibility? I can't help thinking you have an interesting book for a broad audience inside you just waiting to be written." A few tries and several years later, I did finally write that book, and it is that you are reading now.

While I cannot claim the original idea, it seems that I surely must be held accountable in regard to the content; but there, too, I am forced to at least partially demur. For it is also the case that many of the ideas in this book are not mine. I think this is not unusual; in fact, I would say it is the norm. This is of course the point of citing one's sources, as so many ideas we have are but the ideas of others recycled. But my confession is deeper than that. For ideas themselves are sometimes more ephemeral than what one reads and is able to cite. So, here I want to thank all the people I have had conversations with, including many of my colleagues, friends, and former students: the philosophers Andrew Arlig, Andrew Rubner, Nicholas Whittaker, Susanna Schellenberg, Sascha Benjamin Fink, Martin Nitsche, Johannes Brandl, and Carolina Flores and gallerists/art historians including Magda Sawon, Jim Cohan, Flora Brandl, and many others. Conversations are kernels that land on fertile soil, and I am so grateful to those who have helped seed my thoughts by arguing, disagreeing, bantering, supporting, and generally being the gift that we call "friends and colleagues".

Unfortunately, a denial of ownership continues. For the conversations extend much further than that. I have a nonprofit called talkPOPc (Philosophers' Ontological Party club), which is a multimodal project that engages the public in philosophical thought by inviting participants to explore the role of conversation in both art and philosophy and to discover what they themselves think about a particular topic. The project is topic-driven and is based on what I am working on; I do a body of artwork on the topic and write a book. The "Art as Cognition" topic is our third. (I say "our" because there is a team of people who work at talkPOPc, with many ideas being jointly shared and raised.) A large part of the project is holding public philosophy conversations; we have a special tent (and puppet mascot!) that we set up and hold one-to-one conversations between a philosopher (called "Resident Philosopher") and a passerby. We have held hundreds of these conversations (many of them on our podcast) with ordinary, everyday people who have very much influenced my thinking on the topic, that is, this book, and convinced me of the following things. Art affects us because it is a way of experiencing the world through someone else's eyes; it is a kind of epistemic conversation that allows us to empathetically understand one another; and finally, it is the main way we have of escaping the solipsism of our own bodily perceptions.

I remember so many people telling me these things in so many ways: The Caribbean man, a day laborer, who told me that he loved Indian and Chinese music because it took him, so to speak, to different cultures and gave him a different way of seeing the world. Or the homeless man who came into the tent and told me that a particular song will strike him as important because it thereby allows him to understand his own point of view and his own emotions in ways that he had been unable to articulate before. Or the teenaged boy, awkward and guileless, who proceeded to tell me that he loved Jackson Pollock because it was the only time he felt that someone saw him—not only that *he saw* the artist, but also that Pollock's work mirrored so well his own sensibility that he felt he had "been seen," too.

Because talkPOPc re-instills the role of *listening* within conversation, I have come to appreciate more fully the role of listening itself within both art and philosophy or, in other words, the role of conversation. And that has allowed me to repair some misconceptions about both art and philosophy, which has then resulted in this book. For both of these disciplines have traditionally excluded the back and forth of conversation—the exchanges between people—and, instead, have viewed their practices as soliloquies. For example, art is often thought of as the elevated act of high human expression—a pressing-out of emotion—and thus the act of individual genius. Philosophy of art has had an even more wayward journey, for art has often been thought of by philosophers as the handmaiden of Beauty, a triumphant transcending into the stratosphere of an ideal world, again, reifying the artistic act itself. Fortunately, recent philosophy has cut this particular umbilical cord and looked instead toward perception. I, too, follow in this, but I have raised the ante by emphasizing the role of gist perception, as well as sensorial and nonconceptual experience, and examining, then, the communicative functions of those biological facts.

This book lays out how these facts seat the artistic process and how they also form the basis for the perceiver's experience, the connection between the two being an act of conversation. With other perceivers and other conversations, agreement is gradually formed and meaning coheres. To slightly rephrase it, artists are conveying their first-person experience—the "what it's like"—in the artwork. This is a relatively uncontentious claim to artists, although less so to philosophers. Less usual is my contention that much of that "what it's like" is garnered through gist, which is the first 300ms (milliseconds) of perception. Much of this book tracks recent research in visual science on those things and provides the basis for arguing that art is the main way through which we share perspectives with one another.

Thus, art is a kind of conversation that we have with one another. It is a way of handing over to one another the first-person experience of life—the "what it's like" experience. This then allows us to see things from another point of view, to perceive the world with a new attitude, and to learn from one another. It is a way of developing common frameworks, shared perspectives, and social constructs and allows us to remake our shared social reality. Art is an epistemic conversation, which tells us not only how to chunk reality, but also how to care about it.

Therefore, I owe enormous gratitude to all these people, strangers and friends both, who have helped me formulate my thoughts through the absorption of their thoughts.

A small explanation about who would benefit from reading this book. The entirety of this book is directed to upper-level undergraduate, graduate, and professional philosophers, but that is not to say that parts of it are not suitable to audience outside these categories. And while it may not be a book written specifically for "a broad audience"—as the first three chapters lean more heavily on both science and philosophy—the last two chapters of the book would be available to nonspecialists. For example, Chap. 4 analyzes aesthetics from the point of view of the artist, and the culminating Chap. 5 is an analysis of art as a conversation between the artist and the perceiver, which treats related issues of trust, belief, and social cognition. There are also many passages that use particular artworks and detail an analysis of them from an art critic's point of view. These illustrations go far in explaining the more academic points.

And finally, one last denial. In addition to not owning all the ideas herein, the money that supported this project was also not mine. In acknowledgement of that debt, I want to thank several institutions including my own, Brooklyn College (City University of New York), for its support through granting me a sabbatical to work on this book. I also want to thank the Research Foundation at CUNY and the American Philosophical Association for their support through grants as well as the New York City Council Discretionary Funding for their grant to talkPOPc. I especially want to thank Fulbright who sent me to the University of Vienna on a Specialist Scholar award, where I also did valuable research for this project. And of course, I want to thank the editors at Springer, who have been so patient and helpful through this process. I have worked with them before and have found them to be utterly exemplary. Without the support of these institutions, no writing would have happened.

There is though one thing I can take complete responsibility for and cannot lay at anyone else's feet: my children. And to them, the greatest thanks is owed. Without their love and constant support, I do not believe there would be a book. So, to Merida, Dodge, and Arliss: you are absolutely the best life has to offer. Thank you.

Brooklyn, NY, USA Dena Shottenkirk

Contents

Chapter 1
Perceiving: *Gist Perception*

Abstract Gist perception happens in the first approximately 300 ms. One gets a gist of the room before one identifies specific objects; one sees whether an indoor scene is orderly or disordered—i.e., calm or threatening. And whole scene gist takes place in a separate part of the brain than does object-recognition. Why would our bodies have this mechanism and what does it give us? Philosophy has historically focused on object recognition, particularly interpreted as an analysis of how objects become predicable for us. But if we examine the process of gist perception, a different perspective opens up. For gist involves not just passive absorption of the features of the environment, but also an embodied reaction. For example, we experience the first few milliseconds of the fast-moving, charging rhino and feel fear. These affordances are functionally related to the gist experience as it allows for the inductive generalization from low-level features (i.e. illumination, color, texture, etc.) that lead to the semantic high-level category. I argue that our bodies are so constituted as to separate out the public, potentially proprietary objects from privatized, *unnamed* experience. It therefore allows us to think of perception as something other than of entirely linguistically defined objects.

Keywords Gist · Low-level features · Mary Potter · Aude Oliva · Conceptualists · Non-conceptualists · Susanna Siegel · Affordances · J.J. Gibson · Saliency maps

1.1 Introduction

Like knowing how to wink at someone, or how to buckle a seatbelt, *seeing* seems like something we all know about. It's ordinary, it's every-day. It doesn't even take any conscious muscle control—the eyes just open up, look around, and there it is: An accomplished fact!

But the simple and "obviously true" is often a ruse—a scam we play on ourselves. It is one the cruel ways we actually bet against ourselves, hoping that repetition will destroy doubt: "*I know, I know, I know.*"

D. Shottenkirk, *Art as Cognition: How Gist Reframes the Aesthetic Experience as Conversation*, Synthese Library 521,
https://doi.org/10.1007/978-3-032-00143-6_1

But we don't. For perception is the impenetrably complex dressed up in the costume of the simple. And how to remove the costume? It is a speed issue, for thinking often rewards the slow and methodical; it is here where the turtle wins. Once we slow down, look closely, and think hard, we allow ourselves to fall into the subterranean world of precision, where the formerly simplistic and blithe are now clothed in the ever-fracturing minutiae of scientific and philosophical detail. Glorious, precise, fascinating detail that shimmers with truth! And all the way at the bottom of this subterranean journey is gist.

Gist experience is generally defined as holistic perceptual experience within the first 200–300 ms. To give some sense of how quickly this process occurs, we can measure it against the time it takes to blink. Stop reading for a moment and blink: It will have taken you approximately 100–150 ms. Now, think about how much knowledge you could have gained in that time had you been glimpsing something, and you might say "very little". But that would not in fact be true: For despite the fact that this blitz-krieg moment we call "gist" happens in little more than a blink of the eye, it has been shown to give us an enormous, almost unbelievable, amount of information including both high-level e.g., semantic, as well as low-level information.

I've introduced some terms here that need a moment's attention. The distinctions between semantic (or high-level) and low-level are not at all sharp, but the basic definitions of both of these (which are discussed more fully in Sect. 1.3) are as follows: Semantic information is the distinguishing of larger linguistic categories, such as whether it is an outdoor scene or an indoor one, or whether the outlines that are seen conform to a car or a building. Natural kinds would be included in semantic. Thus, semantic information points to larger (e.g., conceptual) categories. On the other hand, low-level features are not necessarily tied to semantic categorization, and thus do not necessarily get tracked by linguistic referents (or concepts), but do include such things as illumination, color, spatial frequency, orientation, hue, contours, texture, motion, and whether it is an orderly scene or a disorderly one. In other words, this is the perception of things not necessarily named, but experienced nonetheless; for example, we can see degrees of illumination but not have words for those degrees and, likewise, can see blobs and vague shapes at low resolution but not have names for those shapes. Thus, when we experience a scene in a gist moment, we reliably know whether it is in an indoor or outdoor scene, its basic objects, and a *very* long list of low-level features.

1.2 Milliseconds

1.2.1 Gist in Seeing and Thinking

But there are complexities here that arise out of the difficulty of actually drawing a line between semantic or high-level features and low-level ones, and, in addition, there are complexities that arise from the fact that we also have some qualitative, subjective responses to that putatively objective information that is coming to us from the empirical, external world. The latter are often called "affordances", and include such responses as repulsion, desire, fear, etc. Therefore, gist, despite being very quickly perceived and giving us an enormous amount of information, is thus also enormously complex. And this makes the analysis of it something that requires not only precision when reviewing specific data, but also, and importantly, the ability to pull back and envision a comprehensive framework for the role gist is playing in our perceptual experiences and our cognitive processes. Where we draw the line between the perceptual experiences of low-level non-linguistic perceptions (e.g., the indistinct blobbiness of the shape) and high-level semantic ones (e.g., it is a triangle) causes many sorts of confusions as does the problem of how (or whether) to divide objective from subjective experience.

It is helpful to take a look at the history of vision science, and its relationship to philosophy. First the vision science. Beginning primarily in the early 1970s when the pioneer in this field, Mary Potter, first began her investigations, the amount of research within vision science devoted to gist is prodigious. Philosophers too began to take note of this research by the end of the twentieth century, weighing in on the implications, and thus though there is both a great deal of empirical, scientific research into gist and some philosophical accounts of its role, it is true that much still remains to be sorted. Some of the problems are evident from the beginning. To take an example from Potter's early research. Much of her early experiments attempted to determine the smallest amount of time required in order to detect and remember a scene (Potter, 1976). Potter noted that is it quite easy for us to recognize and remember a scene that is flashed before us for 1 s, but that at 300 ms about half the scenes are not remembered. The question that she addressed at that time was whether they were, in fact, initially detected and recognized, but that the shortened time restricted their ability to be stored in short-term memory. The experiments consisted of flashing before participants a series of 16 color photographs at 113, 167, 250, or 333 ms. Potter found that there was a difference between settings where the participants were (1) told to search for a target picture—either having been shown the target picture in advance or given the verbal description of the target picture e.g. "a baby reaching for a butterfly"—and (2) the experience, on the other hand, wherein the participant was shown the pictures and then tested a few minutes later for memory. Searching for the target picture gave a success rate of more than 60% at a rate of 113 ms per picture, whereas the group that had the images flashed and then were tested for their memory of them had only less than 11% success rate. (Memory was tested by a recognition test given immediately after the sequence.) In other words, it's easier to detect a target image that is flashed at 113 ms rather than to be able to remember one. The "detection-memory disparity" was important for Potter because it showed, according to her, something about the competing roles of conceptual/cognitive processing and visual/perceptual processing. To quote Potter:

> The present findings thus support the hypothesis that a picture or scene is identified rapidly, whether or not it is expected. Once it has been identified it can be compared immediately with a target description in memory. For a period of several hundred milliseconds after initial identification, however, the memory trace is subject to interference from conceptual processing such as that required to identify a following picture. I have called that interference conceptual masking to distinguish it from visual masking, which interferes with the visual processing that precedes identification (Potter, 1976, 509).

Potter's point, to rephrase, is based on the theoretical positing of a strict and fundamental distinction between conceptual (or cognitive) processes and visual (or perceptual processes). But is she right?

It is easy to accept the notion that we see and understand more than we are able to actively recall or commit to memory. That seems uncontentious. But is the process of committing to memory equivalent to something we would call "cognitive" or "conceptual"? In other words, is the uncontentious claim that we can recognize more than we can commit to memory the same thing as saying there is a radical divide between the conceptual/cognitive and the perceptual? I will argue in this book that it is not and that there is not.

To now introduce some philosophical debates in this regard. Within the field of recent philosophy, opponents have lined up and the opposition teams fall out as follows. There are the conceptualists versus the non-conceptualists, and the liberals versus the conservatives.[1] To take the former opposition first, which rides on the putative division between perception and cognition. This is the traditional view: There is the act of perception and then there is the process of cognition, and the two procedures are mutually exclusive of one another. In this view, perception provides the basic fodder for thought, but thought *shouldn't* assist in the construction of perception (e.g., to think you see something you don't in fact see is a hallucination and thus an instance of failed perception), and perception is simply *unable* to assist in thought—sensory experience is restricted to the embodied realm. Objects are to be seen in perception, and then thought is to use that data in abstraction and concept formation. Moreover, conceptualists (Burge, 2010; Block, 2014) maintain that perception is an act that requires an already conceptualized representation, so that to see an apple is an instance of perceiving an individual instantiation of the larger (conceptual) category; it is that that gives perceivers the high-level features which then grants the perceivers their ability to identify objects as F of x.

Non-conceptualists (Peacocke, 1992; Crane, 1992) on the other hand, argue that perception gives us not already conceptualized but, instead, *perceptual* percepts that must be categorized (and conceptualized) after they are perceived. A topic debated

[1] I give a more complete account of the philosophical debates in Sect. 2.2, but in order to understand the basic disputes regarding gist itself, it is necessary to see the basic disputes in the related philosophy.

widely in the two decades after Gareth Evans' *The Varieties of Reference* (1982), the general idea behind the notion of the non-conceptual is that perceptual experiences are more "fine-grained" than are concepts, and that our ability to experience and represent the world cannot thus be governed by only the application of concepts. Evans wrote only a small section on this, but it has been oft quoted and referred to, particularly the following: "…the senses yield non-conceptual information, whereas language embodies conceptual information…" (Evans, 1982, 123 fn). Evans' view is that objects are not only perceived in terms of being an instance of a named category e.g., "apple", but instead proposed that part of the information that we receive from perceptual experience comes from sensing, and that *that* is non-semantic. What exactly non-semantic, sensed experience *is* is difficult to articulate and many have criticized Evans for not being more explicit. The topic is fully discussed in Sect. 2.2.2, but for present purposes I will note that it seems reasonable to want to argue that lived experienced is not just a matching game where the particularities of experience fall in line under the command of concepts; in other words, our experience is not just governed by some authoritative traffic cops we call "concepts", directing incoming percepts to their appropriate lane. Much of what this book is about is articulating a different option.

The problem though should not be underestimated, nor the current philosophical positions be unduly ridiculed; that is not my intent. The struggle to properly frame and model the problem of perception is one both scientists and philosophers are struggling with, and the opposition between the philosophers labeled "conservatives" and "liberals" shows us again how difficult the problem is. This opposition, too, divides over the putative division between perception and cognition, but this time it is seen in the (related) division between the role of low-level features and high-level/semantic ones. Dubbed the debate over the "admissible contents" of perception, there are those (Dretske, 2015, Prinz, 2013) who argue that perception is only of low-level features and they are called (for reasons that escape me) "conservatives". Their opposition (Peacocke, 1983; Siegel, 2011), called the "liberals", maintain that the kinds of properties that can be given in perceptual experience would include "being an apple". In this, we not only experience shapes and colors, but we also have visual experiences that represent properties that transcend such low-level features, reaching all the way up to the representation of natural kinds. In the oft repeated example of Susanna Siegel's, we not only can distinguish a pine tree from a tree in general, but we then see, in fact, the pine—as it were—in the tree: "Gaining this recognitional disposition is reflected in a phenomenological difference between the visual experiences you had before and after the recognitional disposition was fully developed" (Siegel, 2006, 491). In other words, we can see transcendent kinds/categories.

These two sets of debaters show us something important i.e., that the point of contention lies between the notion of perception and that of cognition: Where (and how?) do we go from the moment of perceiving to the moment of thinking? This is where the trouble lies. I am arguing that it is an examination of gist that reveals the confusions within these debates, and points to a solution.

1.2.2 Gist Is Different from Object Recognition

Now that we have established the terrain of trouble, let us turn back to a fuller account of gist. First of all, and importantly, gist is different from object recognition. While object-recognition takes much longer and requires higher spatial frequency, we see global, holistic scenes very quickly and at much lower frequencies (Oliva & Torralba, 2006). To quote the researchers Greene & Oliva, "Although the object-centered approach has been the keystone of formal and computational approaches to scene understanding for the past 30 years, research in visual cognition has posed challenges to this view, particularly when it comes to explaining the early stages of visual processing and our ability to recognize novel scenes in a single glance" (Greene & Oliva, 2009b, 139).

What is it to see "scenes in a single glace"—what does global perception mean? And how do we experience it? First of all, some perceptual data is reliably obtained at lower time frames. For example, at 50 ms subjects were able to reliably discern *unity* or *order* after only a single glance (Schwabe et al., 2018). Imagine again a blink: the first few milliseconds of perceiving a scene gives you a "sense", as it were, of the setting. Was it a crazy scene? Were you afraid? Or did it seem orderly? Research has repeatedly confirmed that our bodies are quite good at assessing this particular facet of the scene. In other words, order is a quality that is assessed for early.

Think for a moment about how this judgement is accomplished: we instantly see the display of horizontals, verticals, grids, stable patterns, etc. For example, if a tree is before us at a (dangerous) diagonal level, it tells us the level of order. And it tells us to be afraid or at least wary—something might be falling. We see the same thing in visual art: Diagonals in the composition elicit reactions in the viewer of instability (or its not so distant cousin excitement), whereas a completely horizontal and vertical composition conveys orderliness and calm. Not to champion the obvious, but Buddha is never depicted on a diagonal. This is true in real life too: If we enter a room where everything is verticals and horizontal, we feel safe. But diagonals spell instability—the picture hanging crookedly, the table off kilter, items strewn in disorder—it spells discomfort at the minimum and alarm as it progresses.

For this reason, we see and judge orderliness first. It is important to us. To value something on that basis probably relates to basic safety considerations e.g., an orderly environment is in contradistinction to a disordered one, which is often a source of threat. And it is a stable judgement; research has shown we don't often change our minds about this first assessment. Much has properly been made of the evolutionary advantages of gist, and that is obvious: We want some perceptual processes to be efficiently captured very quickly as we are willing to trade accuracy for speed. We don't need to know if the tree falling toward us is a pine or a walnut, or to know that the thing coming at us is a rhino or a tapir; all we need to know is that it is coming. It is fine to trade a number of false positives for a longer life.

Thus, gist gives us an enormous amount very early in the process. Here are a few other things it gives. At 42 ms we can perceive colors (Castelhano & Henderson,

2008, 660); at 100 ms perceivers can tell whether it's a street scene with vertical blocks on both sides (Oliva & Torralba, 2006). We can also tell as well if there are predictably certain kinds of objects such as cars. This is not the same thing as actually picking out the object *of a* car, but instead in viewing the low-resolution image it is recognizing the semantic category that the scene fits into and thus predicting what else might be in the scene. In other words, at 100 ms an observer can say "street scene" as opposed to "indoor scene" and thus be willing to predict that the low-resolution shapes in the scene were probably cars. Thus, this early research showed that 100 ms gives the observer the time needed to identify a complex scene in the form of a picture. If one extends gist to include 300 ms, research has shown that the observer would be able fix that picture in long-term memory (Potter, 1976, 509).

As gist is that experience that happens in that first glance that we have of a situation, it is useful to stop and ask ourselves "why?" Why would our bodies have this mechanism as an essential part of the process of perception and what is it that it gives us?

The reasons quickly become evident once one considers its function and the moments of its occurrence. It is activated in various situations: When one enters a room one gets a gist of the room before one identifies specific objects, when one is presented with a vista of the outdoor landscape one sees the layout of the scene and whether it is calm or exciting, or when one is briefly shown a movie trailer one can recognize the owners of the faces. And though this gist research is only decades old, it is still true that much research has been done on the surprising amount of data that observers can gather in the 200–300 ms that constitute a "glance". It is an important part of the perceptual process. It gives us our initial reaction to the basic data. It frames our response; it sets the stage for further reactions and/or further investigations.

But to think how it functions and to know the time frames involved is only part of the story. It is not only that gist perception occurs at low milliseconds or that it occurs within lower frequencies than does object recognition, but it is crucial to understand that it also takes place in a separate part of the brain than does object-recognition (Epstein & Kanwisher, 1998). The history of this, again, is quite recent. The researchers Epstein and Kanwisher published the first fMRI study on the brain locale of gist in a 1998 paper. That research studied subjects who were shown pictures of scenes and pictures of non-scene objects. It showed that scenes activated different parts of the brain, and that there were two different areas that did so—an area called parahippocampal place area (PPA), along with the retrosplenial complex (RSC). These two were responsible for the scene recognition, but not for the object recognition (Bar & Aminoff, 2003; Kanwisher, 2001; MacEvoy & Epstein, 2011). To roughly locate these areas, the PPA is a sub-region of the parahippocampal cortex and is in the posterior medial temporal lobe that surrounds the hippocampus, which in turn is located in the ventral medial surface—a place close to the base of the skull, anterior to the occipital lobe and posterior to the frontal lobe. The RSC is immediately behind the splenium, which is part of the corpus callosum. This means that the RSC is deep in the midline part of the brain beneath the medial surface of the brain; thus, the term "retrosplenial" refers to the RSC's position behind the splenium. These regions function quite differently from parts of the brain that are responsive to objects. For example, the parahippocampal region responds very poorly to inanimate objects, faces, and bodies, but very well to indoor scenes, outdoor scenes, and relevant functional landmarks such as houses. The RSC has been thought to primarily contribute to scene information in that it allows perceivers to orient themselves in context and plays a role in episodic memory (Maddock, 1999).

Further research has differentiated the specific functions of these two parts of the brain when subjects experience scene gist (Park & Chun, 2009), as well as expanding the role of the occipital place area (OPA) to be included in the process of scene-recognition (Çukur et al., 2016). The latter is thought to extract boundary information for visual scenes during navigation (Julian et al., 2016).

This anatomy further shows us that gist is a process that is separate from object recognition. There is thus general agreement by all researchers that gist experience is processed in a different part of the brain than is object recognition. The latter is accomplished in the inferior temporal cortex (DiCarlo et al., 2012). Thus, while early visual research focused on objects at the atomic units of perception, there has been a shift to research on whole scene perception (Greene & Oliva, 2009a, 2009b). As Oliva, states, "There are many interesting properties of a real-world scene that can be defined independently of the objects" (Oliva & Torralba, 2006, 31).

It is thus a distinct way that we parse the data from the external world. It is important to stop for a moment and absorb the import of this. Philosophy has historically been focused on object recognition, particularly interpreted as an analysis of how objects become predicable for us. But if we examine the process of gist perception, a very different perspective opens up.

1.3 How to Divide Low from High?

1.3.1 Affordances (In Low-Level Features)

Gist is not thought of as a description of the phenomenal character of experience as you might see with an analysis of qualia, or with the traditional Humean analysis of indirect perception where one is perceiving, essentially, one's own sensing. Rather it seen in more quantifiable terms in that it is a measure of what in fact is perceived by the viewer: It is tracking the data as it emanates from the external world. In other words, it is a measure of how much of the environment is actually processed in that short milliseconds' time-blink.

This statement though requires some adjustment. For gist involves not just passive absorption of the features of the environment, but also an embodied reaction: We experience the first few milliseconds of the charging rhino and *feel fear*, or we *feel delight* at the gist of the rapidly flicked movie trailer actors. And while it is not an assessment of the viewer's putatively rational judgement as in a Kantian sense whereby the viewer would deem the art object "beautiful", it is an initial response to the scene and a categorizing of it into larger semantic categories in terms of one's base emotional reactions. This is often seen in terms of the very ground-level responses of aversion or desire: I am *afraid* of the rhino and I *like* the actors. In other words, my very quick reaction to the scene—whether it be fear or desire or any other number of less bifurcated reactions—is not only part of the gist perception, it is *functionally* related to the gist experience as a whole in that it *allows for the inductive generalization that leads to the semantic category.* Feeling the fear upon glimpsing the rhino is *not* incidental to *recognizing* it as a rhino.

One is reminded of course of William James' claim that we run from the bear because we feel fear, and not that we feel fear because we run. Exactly how our limbic system[2] organizes our semantic memory is of course not fully known but the elementary point seems uncontentious: It is our initial limbic system's reaction that influences our semantic high-level categorization. This is a point I shall revisit, but for now, let me only note that the distinction of, and the traveling between, low-level and high-level features is, to some extent, the causal consequence of the initial recognition of affordances.

This might strike some as contentious, so let us instead temporarily retreat to a more conservative approach (that not incidentally will, in the end, allow us the insight into why I am traveling down a less well-trod path.) The distinction between high-level and low-level in gist is fundamentally a division between larger semantic categories e.g., indoor or outdoor scene, buildings or fields, etc., and the particularities of low-level features particularly those un-named with any specificity e.g., the blobby shape, the rough texture, the dim illumination, the vertical configurations of blobby shapes, the particular orientation of a line, etc. This is the usual view, and it is not entirely wrong, although it is, I shall argue, incomplete, and thus leads us astray. In this usual view, the role of the high-level features is thought to be quite distinct from those of low-level: With high-level, we are moving vertically up the ladder, so to speak, to greater abstraction when we identify semantic meaning e.g., F is an instance of x. And x is something quite distinct from F! This is a continuation of ancient metaphysics that sees universals and particulars as two different ontological kinds. Thus, linguistic naming/semantic meaning—which are constituents of concept identification—form a separate sub-kind in gist perception in contradistinction to low-level features. They seem to point to ontologically different entities and they seem to be cognitive and mental in a way that the perceptual experiences of low-level do not. At least this is the usual view.

[2] Limbic (from the Latin "border") is between the neocortex and the subcortical structures, and it includes the hippocampal formation, amygdala, cingulate cortex, septal nuclei, entorhinal cortex, perirhinal cortex, and parahippocampal cortex. It is the center for emotions, behavior, memory, and the regulation of stress.

Thus, there seem to be the two issues as hand, which I will name as the "affordances issue" and the "concepts issue".[3] As I have briefly pointed to the first e.g., affordances, at the beginning of this subsection, let me then address the latter.

1.3.2 Concepts (In High-Level Features)

To reiterate the point, the concepts issue shows itself in semantic naming as a kind of categorization that happens, for example, when we respond to a gist experience by identifying it as an instance of a larger category such as an indoor or outdoor scene. We are able to identify the scene and know what semantic category it is an instance of. Let us just grant the example. But now we have a problem. Is this seeing or is this thinking? What exactly is happening in this gist moment when we are able to identify the scene's semantic category—it is perception or is it cognition?

To recapitulate, there are two branches to this mutual exclusivity that have both been held by most. (1) Perception is conscious perception and is immediate, sensorial, constrained by the spatial existence of the objects which it is a perception of and by the causal relationships that govern the physical world. Perception is an openness to the world. (2) Cognition is free from the constraints imposed by perception. Even though some of the building blocks of thought come from the input provided by perception, thought's ability to re-combine, to abstract, to draw inferences and logical relations, to structure categories through the process of forming concepts, etc., sets it apart from perception. It is also the realm that gives us the "I" of identity. We do not locate our identity in our acts of perceiving but in our thoughts of who we are. How then do we know if something is an instance of perception or cognition?

The answer seems to lie, for most, in the line that is drawn between high-level (semantic) and low-level. For many philosophers, semantic meanings embed concepts, and concepts are, depending on your ontological commitments, mental (e.g., abstract objects, often seen in the form of propositional attitudes) or they are (if you are a nominalist) instances of general terms. For scientists, the fine philosophical distinctions are not usually made and instead the default position seems to simply be that they are different kinds of procedures; semantic gist signals computations whereas low-level features are more mechanical, non-computational, non-cognitive, and hence mere perception.

But here is where the hand-waving begins. Almost all writers, in both science and philosophy, decry the ability to clearly distinguish between what we call low-level and high-level. At what point does an un-named blob become a signifier for a

[3] The notion of concepts is a barbed issue, and one that can't fully be dealt with in a book that is not devoted to the subject. From Ryle's arguments against the destructive pedaling of these false posits in his *The Concept of Mind* (1949) to Peacock's trumpeting of their necessity in his *The Study of Concepts* (1992), debates still rage. But while sitting more closely to Ryle's position, I cannot in a book devoted to how art is a kind of cognition fully engage on why I would side more with Ryle.)

lake? Thus, the central problem: W*hen* do we—and *how* to do we—go from the first state of blob-awareness to the next one wherein we are categorizing the shape as a lake? And when do we go from perception to cognition? If seeing the lake necessarily involves the conceptual category "lake", then high-level features are obviously conceptual. At what point does that happen?

Many writers are content to simply say the line is vague, we are unable to draw the line, or to proceed without an explicit distinction or definition, and yet—and yet!—to continue to write and analyze as though the distinctions *were* in hand. It is a treacherous kind of move, as it is assuming exactly what one is not able to establish, and I would argue is the core of much of the problem with gist research. It is, I would argue, very important to be clear about the division between low-level and high-level, and in what context we go from one to the next.

1.3.3 Induction (In Going from Low to High)

Some researchers though are, in fact, a bit clearer about the relationship between low-level features and high-level semantic features. As an example, M.R. Greene and Aude Oliva, prominent researchers in the field of vision science, argue that it is the build-up from the low-level features that then justifies the inductive generalization that is at the heart of the high-level semantic categorization. They use as their example the fact that subjects are more likely to "confuse river and forest scenes, which have very similar spatial layout properties…than to confuse forest and field scenes, which have very different spatial layout properties" (Greene & Oliva, 2010, 1430). In conclusion, they state, "Altogether, these results suggest a global, scene-centered view of scene understanding in which the meaning of a scene can be understood from the rapid computation of global scene properties representing aspects of scene structure and affordance" (ibid.). And in the abstract of a related article they state, "These results provide support for the hypothesis that rapid categorization of natural scenes may not be mediated primarily though objects and parts, but also through global properties of structure and affordance" (Greene & Oliva, 2009b, 137).

To paraphrase and expand on the point: The processing of spatial relations, which is a central function of both RSC and OPA, give forth pattern recognitions which in turn can be matched with larger categories, but—and this is important—those jumps in inductive patterns are made *in consort with the affordances of the scene*. Those affordances begin with elementary responses involved in determining such issues as: Is it spatially open or closed? Is it ordered or disordered? Is it pleasing or frightening? Again, the notion of affordance enters, and again *with a causal and functional role*.

Other related research is often arriving at a similar conclusion. For example, researchers using whole-scalp MEG techniques tested the relationship between neural processing and visual information categorization in gist perception between 100 and 250 ms, finding the "temporal relationship between when low-level scene-related information becomes available and when it becomes useful for behavioral categorization" are co-localized with visual information representation (Ramkumar et al., 2016, 295). One of their main conclusions regarding the data is stated thus, "…rapid-scene categorization is comprised of highly parallel sets of sub-processes. Therefore, neural activity that is relevant to behavioral choice is temporally simultaneous with the encoding of low-level visual features" (Ibid., 300).

To reiterate the point. This particular research was assessing the neural basis of rapid scene categorization, attempting to correlate the neural processing with the behavioral responses. As the researchers stated, "we aimed to distinguish between brain regions that are involved in primarily representing low-level visual information that is captured by the spatial envelope and brain regions that are involved in representing alternate visual features, or potentially in making judgments about the scene category, as represented by behavior" (Ibid., 299). In other words, it is the relationship between low-level information and high-level categorization processes. The question was whether or not scene category selective information could be explained by the processing of low-level information. In their affirming conclusion they state, "yes".

In their quote from two paragraphs previously, the pertinent phrase here is "relevant to behavior choice". Thus, our neural processing of the inputted information that becomes useful for behavioral categorization might not be limited to linguistic, or concept, identification. As William Fish has noted, "Although language is a way in which we can treat square things differently from nonsquare things, it is not the only way. General behavioral capacities provide another possibility" (Fish, 2021). He's right. We navigate the world, choose one thing over another, register dissent, cooperate with others, etc., all quite apart from articulating or even forming relevant linguistic, semantic categories. Capacities to act are surely a way to frame the problem independently of the traffic-cop of concepts, as are partial credences of belief. The emphasis here is on acting.[4]

Let us look again at the issue that I am calling the "affordances issue", and perhaps the glimmer of a solution will appear to the problems of both how gist controls further perceptions as well as how to divide high level from low level. Note again the connection of the PPA area to the amygdala and the role in the larger limbic system e.g., the center for integrating our emotions.[5] If one thinks of flipping rapidly through a book on the history of art, for example—a perfect instance of gist perception—each page will elicit an immediate evaluative response: Not 'good' or 'bad' but the underlying emotional states of attraction or repulsion. *We move toward or away.*

[4] Capacities to act, a notion we originally inherited from Aristotle, has gotten a recent infusion of support from philosophers such as Susanna Schellenberg in her *Unity of Perception* (2018). Also relevant is Lara Busak and her notion of credences of belief in the paper "Belief, Credence, and Norms" (2014) and Sarah Moss' *Probabilistic Knowledge* (2018).

[5] I use the word "emotion" with great reservation; I believe that it has such a large weight of associative meanings that we cannot get to the proper reference of the term. And, so, I will use it as little as is possible, substituting instead more specific and precise language.

Thus, much of gist recognition is not on the conscious level. But our bodies are registering the affordances provided by the gist stimulus. Researchers have noted the role of the amygdala in this, particularly in regard to allowing us to register the difference between something we care about and those other things we do not. "A critical function of the amygdala appears to be to segregate the neural representations of the significant from the mundane…the amygdala is believed to shape perceptual experience" (Lim et al., 2009, 16841).

This is the basic beginning for action: Am I going to move toward this or move away from this? The issue of early gist recognition regarding the level of order/disorder is intimately connected to this, as are other very early low-level responses to open/closed spaces, etc. In fact, it is the latter that has garnered the attention of some researchers, who have concluded that, despite the fact that "…studies have argued that these regions represent semantic or categorical properties of scene…in high-level cortices…[but] it does not necessarily follow that patterns of neural response are systematically organized by semantic or categorical properties, or that the perception of these properties is causally linked to such patterns. Indeed, other studies have suggested that responses in these regions may be better explained in terms of the spatial characteristics of the scene such as openness, or distance" (Watson et al., 2017, 1).

In conclusion, low-level features have profound effects. It is reasonable to say that we start in a state of inputting the data from the environment, and very instantaneously we register our affordances to that input in the (admittedly) over-simplified bifurcation of attraction versus repulsion (there are of course many gray areas between the two). We then continue to refine that input, allowing ourselves behavioral actions e.g., such as looking more by absorbing more low-level features (in the desire instance), or pulling away (in the repulsion one). There are also, *potentially*, inductive generalizations into larger semantic categories; an example would be where the property of being open-spaced is now attributed to the semantic object of outdoor scene. (More about property attribution is discussed in Sect. 2.4.2).

But two points need to be made here: (1) the latter fact proves that there are two independent steps. One can experience the low-level features of open-spaced without moving to the larger semantic category and thereby identifying the scene *as an instance of* the category "open-spaced". This is very important in perception in general and will be even more important when we consider the perception of artworks. To reinforce this fact, we need only to refer back to Greene & Oliva's research on how mistakes happen in the inductive process toward semantic generalizations. The fact that mistakes are made in moving to the semantic category from the low-level features *proves the division of the two processes and establishes the low-level as independent and non-semantic and hence non-conceptual.* This, of course, serves as an argument against the conceptualists, who want to argue that perceptions are already conceptualized. But neither does it prove the position of the non-conceptualists, for in order to prove that you would need to also establish that the inputs were entirely feed-forward as non-conceptualists want to argue that the perceptual inputs (to cognition) are themselves non-cognitive and that cognition happens *after* the perceptual inputs are in place. (2) Furthermore, and importantly, it additionally proves that low-level can be processed independently of high-level features, and that that independent processing gives low-level features a degree of autonomy and a separate role to play. Thus, it is not a necessary posit that all low-level features are experienced in the service of linguistic semantic categories. We perceive low-level gist and then, perhaps,—*but not necessarily*—linguistically generalize the information into larger concept categories.

Now we have established that low and high level features are separable in gist, the point that remains is how to properly characterize the two categories of low and high. Historically, it has often been concluded that the division between low-level and high-level features can be seen at the level of initial neural processing restricted by both time frames and the specific brain regions devoted to global scene perception. In other words, and in regard to the first point, the initial few milliseconds were more likely to be devoted to low-level features, while later times were devoted to the top-down mechanisms of high-level. But recent research has challenged that view, at least to some extent, with considerations of the non-sequential and interdependent nature of low and high level features, the complex and dynamic feedback systems involving many different parts of the brain, the role of probabilistic calculations accomplished via peripheral vision in contradistinction to an analysis of foveal vision, etc. All of these points will be covered later in the book, but suffice it to say for now, research is still being done on exactly how we process differently low-level and high-level, although we can still flag up the obvious fact that it is the linguistic fact of naming that is centrally involved when we go from low to high. And this linguistic fact is underwritten by our ontological parsing of the world. The division can also be seen as well in the non-linguistic nature of those elemental, un-named particularities of low-level features. Furthermore, it is also important to remember that the leap from low-level to high-level is an inductive one, with all the benefits and pitfalls associated with that process. This is a point we will repeatedly return to.

Now to look at the philosophical implications of this. To reiterate, for conceptualists, perception gives us an already conceptualized representation, and that is highly dependent on object perception, which itself is governable by general terms. For example, we see the low-level red blob in front of us an "apple" (a high-level feature) because we have the general term (e.g., the concept) of "apple" and we are seeing the red blob as an instance of the general term. For non-conceptualists, the general term will still claim ownership of the blob, it just does it at a later date. For most vision researchers, the division rides on a prior (and equally problematic) division between (mechanical) perception and (computational) cognition.

But we ought to first ask "what is the function of their division?" I would argue that the division between low and high feature in gist doesn't ride on the characteristics of mechanical versus computational or on the division between conceptual or non-conceptual, but on a more obvious characteristic, and that is this. It is to separate out the public, potentially proprietary objects from the private, non-proprietary experience. I can't own or pass over my ownership of my private phenomenal experiences—e.g., my experience of that blob in front of me. But I can have a proprietary with the boat. Furthermore, and importantly, *we can be wrong about semantic, high-level gist in ways we cannot be wrong about low-level gist.*

We *go from* low-level to high-level gist, taking as veracity the low-level data. This is a central point. *The semantic category is provisional, apt to be re-evaluated, re-defined. It is an arena more laden with mistakes and minefields. And it is not necessary for many perceptual experiences.*

If this is correct, then semantic categories lose their causal efficacy vis a vis the conceptualists' story. While is it clearly true that we are able to identify a particular thing when told to find an instance of that larger semantic category e.g., when told to identify a boat in the target image we are in fact able to see the boat (see Potter's original research), this does not mean that *when we start with the individual low-level particularity* that we are only able to process it *if* we see it an instance of the larger category. Going in one direction doesn't entail the necessity of the other direction.

To slightly rephrase and clarify: if we are told by the researcher that we are looking for a boat, and then have flashed various pictures one of which would be a boat, and then are able to identify the flashed boat picture, we are going from the directive of the researcher's semantic knowledge that encapsulates the general term/concept and traveling down to the low-level features that are experienced in the flashing of the image (e.g., the blob shape), and then up again to the category itself; a match was made. But that correlation between high-level and low-level is not entailed when we go the other way. Think about it like this: If we start with the low-level experience we do not so easily travel to the correct semantic category, or in fact, necessarily to any semantic category. We can be presented with the same blob shape in a low-level experience such as when we walk down a street and quickly glance in a direction and see that blob; but we mistake it for a car instead of a boat in front of a house. It is hard to do the inductive generalizations when we start with low-level, but starting with the semantic category and then matching perceptions is easier.

Thus, the difficulty with the conceptualists' account that wants to maintain the entailment is as follows. To reiterate: The conceptualists' claim is that perception is the process of identifying low-level features as instances of high-level semantic categories; that is, we are able to perceive the individual dog because we recognize it as an instance of "dog". Thus, the philosophical position of the conceptualists (1) reasserts the dominance of an object-centered world, (2) ignores the roles of perceptual particularities, and (3) ignores general behavioral capacities as well as behavioral actions. Even more, it ignores the logical fact that there is no entailment of a necessary correlation between the low-level and high-level when we start with the low-level. The x can just be an x.[6] Thus, the question then becomes whether concepts govern the process or are they merely a non-necessary and only sometimes employed mnemonic device? To rephrase: Are they, perhaps, just a kind of perceptual shortcut on the keyboard of perception, possible but not necessary?

[6] Sometimes a cigar is just a cigar.

1.4 Role of Low-Level Features

1.4.1 The Veracity of Low-Level Features

If we assume the plausibility of my claim that high-level semantic categories distinguish themselves from low-level primarily in their statistical frequencies of unreliability, then we have a view of the first 300 ms that constitutes gist as one that is giving us reliable rapid fire low-level information, which is the basis upon which we employ inductive generalizations that in turn form semantic categorization. The latter of course are sometimes subject to mistakes and hence revisions: We are likely to "confuse river and forest scenes"—mistakes happen.

The question is now to ascertain how the data that is processed in gist affects or guides the rest of our perceptual functions, particularly in regard to the role of low-level features in those first 300 ms. I will first examine in detail exactly what constitutes low-level features, then discuss the role of saliency maps with low-level features, and finally give narrative examples of how we experience the role of low-level features. As of now, I have not mentioned art. That seems like a bit of a shortcoming in a book devoted to it. So, at the end of this subsection, I will use examples from art to illustrate how low-level features guide the rest of our perceptual functions i.e., from the point of view of the perspective of the viewer.

We have seen already the fact that many low-level features can be detected at <100 ms, such as color, which can be detected at 42 ms (Castelhano & Henderson, 2008, 660); and at the threshold of 19–67 ms (reaching asymptote at 100 ms) subjects were able to classify natural versus manmade more quickly than they could make a semantic classification of the scene: "Remarkably, the threshold presentation duration was, on average, shorter for perception of a scene's global properties than for perception of the scene's basic-level category" (Greene & Oliva, 2009a, 471). This shouldn't surprise us. The low-level features of the scene's structure contribute to the global properties of gist perception (Ibid.). There are going to be fewer mistakes at the very basic discriminatory levels such as man-made or natural as the inductive optional inferences are fewer. This is a simple issue of probability. But what really constitutes the basic level low-level features?

It is useful here to reiterate the types of low-level features: illumination, color, spatial frequency, orientation, hue, contours, texture, motion, and whether it is an orderly scene or a disorderly one. Again, these are not necessarily tied to semantic categories, although in some instances—in a given context—they could be. For example, one might see a yellow in the first 42 ms, and if one were queried later about the color, would be able to name it. It doesn't mean though that the semantic category "yellow" consciously accompanied the perception of the yellow particular itself. While test subjects were able to identify their perceptions *after* the perceptions themselves occur in the instances of high-level features, e.g., seeing the boat after being told that the target to identify is "boat", categorizing under a general term is not always the case with the perception of low-level features, most of which are without names or semantic category. This is worth reflecting on for a moment.

We are used to thinking perhaps of perception as being only conscious and of named entities, but that is not true. (We are used to thinking like conceptualists!) For example, the particular blobby-like contours of a scene will be seen at low spatial frequencies and at, as well, <100 ms, but that particular shape need not be identified with a name, or, in fact, even have a name. Nevertheless, despite the fact that we don't name these things, we are able to perceive this information and use this in our decision-making processes, allowing us to act or refrain from acting. As research has shown, the brain exhibits "strong sensitivity to low-level summary statistics…100 ms after stimulus onset confirms that this information is available early in processing" (Groen et al., 2013, 18821). Low-level features are the central, and yet often neglected, constituent of perception.

Let us take a moment and examine the related notion of low spatial frequencies, which will allow us a fuller comprehension of the role low-level features play in our experience. Spatial frequency is based on a unit of measurement from the retina to the thing perceived and is a measure of the perceived temporal and spatial changes of that thing. For example, if you were to hold your index finger at your arm's length, the image the nail creates is one-third of a millimeter; and if you drew two lines from either side of your nail to your retina, those two sides would geometrically create an angle that would subtend one degree on the retina.[7] As the object gets bigger e.g., closer, the angle gets bigger. If the object is in very close proximity, such as the fine lines on a person's face, it is an instance of high spatial frequency. If though it is at a greater distance, it is low spatial frequency. It has also been shown that the low-frequency channels are experienced first, thus perceived at very short durations of time.

The processing of low spatial frequencies in short time durations are activated before object recognition. The resolution of an image is often described in terms of vertical bars, where the resolution is measured by the closest spacing of the black and white bars that can be resolved. In this, spatial frequency is the line pairs per millimeter—or the number of wave cycles. It has been shown that "oriented blobs in a particular spatial organization at a resolution as low as 4 cycles per image…would initiate recognition before the identity of the objects was processed" (Oliva & Schyns, 1997; Schyns & Oliva, 1994). Because spatial frequency (which also occurs in sound) is of any object that has duration in time and is across space, and it is determined by how often the sinusoidal function repeat over that distance—a measure of how much the optical image changes in terms of cycles per millimeter—spatial frequencies are how visual information is processed and, in this, the division between low frequencies and high frequencies is important.[8] Spatial frequencies range from very fine grained to very course; some neural processing (e.g., of low

---

[7] See http://www.psy.vanderbilt.edu/courses/hon185/SpatialFrequency/SpatialFrequency.html. The material from the Vanderbilt site was excerpted from *Perception of Visual Motion* by Sekuler et al. (2002).

[8] Researchers are not yet clear regarding the question of where in the brain both low frequencies and high frequencies are processed. See: https://www.frontiersin.org/articles/10.3389/fnint.2014.00037/full

frequencies) will give us the extremely quick and very course spatial frequencies where we see the forest but not the trees, whereas other neural processes (e.g., of high frequencies) give us the trees themselves.

The specifics of this are interesting. Neurons have receptive fields of different sizes, and they register the presence of borders or edges at different scales. "For example, when a short row of retinal ganglion cells is excited, we perceive an edge" (Balasubramanian & Sterling, 2009, 2754). We have evolved in such a way that our vision provides multiple versions of the external world, e.g., multiple "channels", thus allowing us to extract different kinds of information at different moments and for different purposes.

To get a feel for what it means to be aware of low-level features (or information, etc.) that we garner in low-resolution images, we can look at an illustration taken from Aude Oliva's research, where we can see how it is the low-resolution image provides for us low-level features such as color, shape, orientation (often of lines), texture, etc. These are of central importance in determining the structural representation of the scene. The images below show an outdoor city scene which are layered on top of an indoor hallway scene. In (a) the hallway scene is high spatial frequency, 24 cycles per image (HSF) while the city is low spatial frequency, 8 cycles per image (LSF). In (b) it is reversed. The LSF image has only blobs, while the HSF image has details, surfaces, object contours. Here is the interesting trick: If one squints one's eyes, the LSF is able to be immediately perceived, and in fact, people do perceive that at 30 ms, well before the 150 ms of perceiving the HSF (Oliva, 2005, 253) (Fig. 1.1).

The two experiments revealed that hybrids were preferentially categorized in a coarse-before-fine sequence. This is important. The low-level features, given to us in low-resolution spatial frequencies, provide a fundamental, embodied grounding that allows for instantaneous awareness of such essential features as orderliness, one's spatial relationship to the image. In this, the low-resolution indoor scene in (b) makes us aware of being in an enclosed space in contradistinction to (a). It tells us how we stand to that external presence. And there is a set of affordances that are consequent to that. This is an important role of low-level features.[9]

[9] Many kinds of research relies on similar low-resolution and low-level features. For example, the recent decoding of the 2000 year old Herculaneum papyri was partially solved through this. As a wonderful example of a project designed both as a competition and as a cooperative venture, one scientist, an astrophysicist, used a technique he called "persistent direct visual inspection" to decipher some of these scrolls. His blog post on that is https://caseyhandmer.wordpress.com/2023/08/05/reading-ancient-scrolls/. It is my contention in this book that part of the reason humans have visual art is that it is a way of practicing this sort of deciphering.

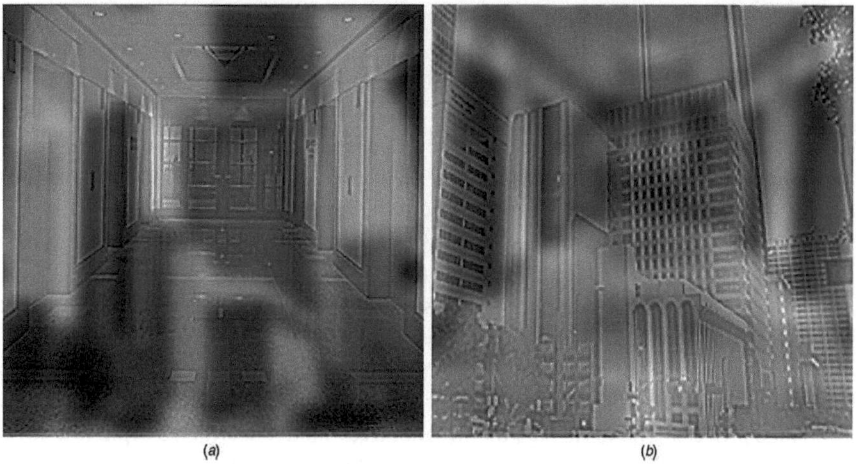

Fig. 1.1 (a) A hybrid image representing a hallway scene in high spatial frequency (HSF, 24 cycles per image) and a city scene in low spatial frequency (LSF, 8 cycles per image). If you squint, blink, or defocus, the city scene should replace the hallway scene. (If this demonstration fails, step back from the image until your perception changes.) (b) The complementary hybrid scene, showing a city scene in HSF and the hallway scene in LSF. (From Schyns & Oliva, 1994; Oliva & Schyns, 1997)

1.4.2 Saliency Maps and Editing

These low-level features are constitutive of the mechanism that absorbs the external world through sampled episodes driven by extremely fast saccadic scans. These saccadic scans happen at three to four per second. The usual view is that while these are mechanical, computational samplings, they are also highly edited; we do not process all that is perceptually available (Tatler et al., 2005, 643). This is an important point. If you remember the role of the limbic system in perception, particularly the amygdala, and what I am calling the "affordances issue", then this is not surprising. While there is disagreement (of course!) among researchers and philosophers regarding the exact roles of top-down and bottom-up features, there is growing evidence that our perception is highly edited through a function largely governed by so-called "saliency maps."[10] These are "maps" which determine what is salient, and what is worth ignoring (see Lim et al., 2009). These saliency maps are, in turn, significantly constituted by low-level features: "This model consequently represents a complete account of bottom-up saliency and does not require any top-down guidance to shift attention" (Itti et al., 1998, 1254). The researchers Torralba, Oliva, and Castelhano explain saliency maps as a selecting of the significant perceptions into usable data. Granting computational efficiency, "Saliency is independent of the

[10] I will briefly explain in this section as well the distinction between preattentive and attentive, and the fact of iteration, but will elaborate more fully on all of these in subsequent chapters.

nature of the particular task, operates very rapidly and is primarily driven in a bottom-up manner…" (Torralba et al., 2006, 783). This combines with targeting provided by context and Bayesian models, giving an attentional "mechanism driven by image saliency and contextual guidance" (ibid.). This system if based on automatic calculations of probability, such that they that are "a natural consequence of the probabilistic framework," which allows us to identify global features as well as subsequent objects (ibid.). Probability and induction are again writ large.

Exactly how does this process occur? Because these sampled episodes of fast saccadic scans occur at 3–4/s; this makes for the obvious fact that a large proportion of our visual surroundings are not processed by us (see Ballard et al., 1992; Land & Hayhoe, 2001; Land et al., 1999; as cited Tatler et al., 2005, 643). What distinguishes the saccade target selections from those that are not targeted? Beginning in the 1980s, many studies began to appear that characterized visual perception as a two-stage process: The first stage of "preattentive" mode, in which simple low-level features were rapidly processed in parallel over the entire field, and a second "attentive" mode, where the focus was to particular locations in the visual field (Koch & Ullman, 1987, 115). While there is agreement that the process involves both top-down and bottom-up guidance factors, there are a significant number of researchers who have emphasized the bottom-up role of "salience maps", and the activity of low-level features in those salience determinations (Itti & Koch, 2000; Itti et al., 1998; Olshausen et al., 1993; Parkhurst et al., 2002). Granting computational efficiency, "Saliency is independent of the nature of the particular task, operates very rapidly and is primarily driven in a bottom-up manner…" (Itti & Koch, 2000, 1490).

When determining what the oculomotor system prioritizes as salient, researchers have found that "neuronal responses are…strongly modulated by context" (Itti & Koch, 2000, 197). In other words, our initial scanning gives us data that contrasts with its surroundings, so a single red shape will show up in a field of green or a diagonal in a field of straight lines. This gives us a model of vision that is initiated by a bottom-up oculomotor system governed by relatively uniform standard of saliency that is, then, in both a feedforward and feedback loop, interactive with top-down object-recognition processes. There is a strong presumption that the editing also takes places at a sub-object level where the selection of low-level features would conform to inductive pattern generalizations i.e., a habitual preference for a certain color would edit for that color at early low-level perceptions making it more salient for a particular perceiver. This has obvious evolutionary advantages, as we need an interpretative map of the world that can quickly prune incoming sensory input, and the bottom-up processes do exactly that. It also provides a role for decision and top-down cognitive phenomenology.

Research has also shown the process to be iterative. "Itti et al. consider a purely bottom-up model, in which spatial competition for saliency is directly modelled after non-classical surround modulation effects. The model uses an iterative spatial competition scheme with early termination. At each iteration, a feature map receives additional inputs from the convolution of itself by a large difference-of-Gaussians filter" (Ibid., 198–199). Choices must be made, and those choices must be sequential, or the saliency map would be as cluttered and difficult to interpret as the original scene; thus, the process is iterative. And, of course, all this is done very quickly. Speed is important here, which again speaks in favor of the important role of global scanning as low-level processing occurs in the range of 25–50 ms per item, whereas the top-down attention (most probably controlled by higher levels in the brain including the frontal lobes) takes around 200 ms or more (Ibid., 194). It is important to note that recent research does not view this reliance on low-level information as a feedforward process, as was originally thought by early researchers, but that significant amounts of low-level information function in a feedback loop (Herzog et al., 2016, 11).[11] This further emphasizes the role of low-level features in perception.

Gist, as global whole-scene perception, then, is significantly the result of processing low-level features. As one of seminal researchers in the field, Aude Oliva, states, "…the global features are originally built as combinations of local low-level filters of the type found in early visual areas" (Oliva & Torralba, 2006, 34). Since low-level features play a significant role in gist perception and thus perception itself, and, in addition, as they are experienced independently of object recognition, it therefore allows us to think of perception as something other than of entirely linguistically defined objects. In other words, as an empirical analysis of the role of low-level features, gist research allows us to step back from viewing percepts as entirely constitutive of objects. Therefore, it allows us to focus on the entities (however we might metaphysically define them) that are constitutive of those low-level features of sensory perception, despite them being unnamed. And this, in turn, brings us, once again, to the consideration of perception conceived of as something other than linguistically identified objects. This is made obviously true as gist is whole-scene experience and not of discrete objects.[12]

1.4.3 Art: Low-Level and Saliency

Let us now use some examples in art. What do I see when I look at an artwork? And *how* do I see it? In particular, what role are low-level features playing in that story, and how do the initial gist perceptions frame the subsequent ones?

I'm going to use a historical painting for this exercise, not a contemporary one. This is, needless to say, a non-laboratory experience, including both what I (subjectively) take to be the first brief gist experience of the work. This is followed by the many multiple takes of the perceptual experience as I repeatedly re-look at the work.

[11] For a complete explication of this see Sect. 3.3.

[12] See a related discussion of this in Eric Mandelbaum: "That said, it might still be reasonable to suppose that one could use questions of categorization for illuminating the debates about the contents of experience. Insofar as one takes perceptual experience to be determined by perceptual processing, the discussion here will inform the possible contents of perpetual experience, greatly shrinking the allowable property of high-level theories like Bayne and Siegel, while enlarging the contents of low-level theorists' ontology (Mandelbaum, 2018, footnote 14)".

It is in that way a reporting on what happens to us in ordinary everyday life as we experience the world, particularly when we look at an artwork. For it is in the viewing of an artwork where it is most evident how the initial gist of the scene takes us in or fails to take us in, and how the subsequent repeated lookings follow in that trail. Thus, what I am about to describe is a re-telling of the actual process of how we look: How the first glance directs us, how we confirm that first gist experience particularly that of "order". But it is also a telling of how in some important instances we can, at later stages, absorb different perceptual data that contradicts and denies *some* of the first judgments, making for a perceptual experience that yields more "complexity". I will argue that the original and first experience of order (vs. disorder) is the guiding one. The entire perceptual experience was timed at 6 min. But note the obvious please—while the initial gist experience would count as the first 300 ms, the rest of the saccadic scanning and longer fixations link causally to those first 300 ms. Furthermore, each of those separate scans are themselves gist perceptions. Thus, the whole 6 min is seen as a whole, governed by gist.

I walk into the Metropolitan Museum, make my way up the wide, stone front staircase, and go into gallery 625 (Fig. 1.2).[13]

It is a large-seeming room, with the usual museum lighting that is neither dim nor bright, and there a about two dozen paintings hanging on the wall. And when I look to my right, I see a row of Velázquez paintings hanging. I walk over to the middle one.[14] What do I see in my first micro-glances?

The painting is of average size and relatively somber in tone, mostly monochromatic, with only two lighter colored shapes that are the most obvious contrasting hue, one smaller one and one larger shape. This larger white is the main thing that registers. Both of them are off-center to the left, giving the painting a slightly off-kilter and somewhat edgy feel, and both of them are part of a larger shape (still off-center and on the left). This shape—the only one in the painting—is a straight up and down vertical one, which feels weightier as it is on the left-hand side; it makes the painting almost tip to the left. The composition is thus not symmetrical, given the off-kilter composition. The overall color is a tan and greenish. This is what I would have seen on my first gist.

Looking more closely and allowing the foveating to proceed, I see the face, whose forehead was one of the highlight shapes, but the face is not entirely white as it is of a black man. My eyes are next drawn to the extreme white shape that first caught the contrast neurons' attention; it is a white lace color, attached to what looks like the suit of a reasonably well-to-do person. Now that my eye is traveling downward, it goes to the next contrast, a less obvious one. For there is also a hand that I now notice, which is more to the right side of the painting and at the same distance from centerline than is the face, which is to the left of the centerline. It has highlight on it, so it is possible to pull it out from the surrounding greenish browns. But now

[13] To view this in color and higher resolution, see: https://www.metmuseum.org/art/collection/search/437869

[14] As a disclaimer, I should say that I was formerly an art critic, on staff at both Artforum and Art in America. I have looked at and written about a lot of art, so this is not the experience of a novice.

Fig. 1.2 Valazquez Diego Rodríguez de Silva y Velázquez. Juan de Pareja 1650. Gallery 625. **Dimensions**: 32 × 27 1/2 in. (81.3 × 69.9 cm) (https://images.metmuseum.org/CRDImages/ep/original/DP-14286-001.jpg?_gl=1*14mx4oz*_ga*OTE0ODMzNzI5LjE3MDQ2NDA5MDM.*_ga_Y0W8DGNBTB*MTcwNTk0MTUzNi44LjEuMTcwNTk0MTY3Mi4wLjAuMA)

I notice that the greens there are soft, luscious, like the floor of an evergreen forest. The hand also is next to his stomach, which slightly poofs out, added a soft shape to the soft color. The hand itself and its highlights add some ballast to the composition, but it is still the case that the figure takes place primarily in the left half of the image. But the hand, the face, and a very slightly highlight area of his cloak, as it trails off the left-hand side of the painting, make a perfect triangle. This triangle provides some grounding; his figure then sits in that triangle, but it is almost as if he is leaning back against a wall, or is slightly recoiling from the observer, or as if, perhaps, he has pulled himself completely upright and rigid and therefore has given himself more space in front of himself than behind. In any event, the effect is a solid man with a self-possessed, confident posture. This shows in the face as well, which is absolutely extraordinary. Calm, poised, completely aware, he looks the viewer straight in the eye. There is a tremendous amount of sensitivity in the face; I regret not being able to meet him.

But the dominate shape and color is still the super white collar, which was the main highlight area that I noticed during the initial gist. And now that I've looked harder, I see the function of that. It is like a visual float that holds up the head. The sitter's regal posture and self-possessed demeanor are visually supported by the whiteness of the color. But, as it is also off the left side, it throws the center of the canvas off. This, too, has its function. Because the composition is not static in a grid like way, I approach the painting somewhat more circumspectly; I am not being giving a blatant easy-to-read story, but something that requires more careful leaning in, more nuance. And his face rewarded my more careful leaning in.

Let us now go back and examine the initial gist of this painting, and see how much the initial gist was carried through to the more foveated analysis. First of all, consider the somber tone and the left of center composition. The former stayed consistent, but was slightly contradicted when I looked closely at the colors around his hand. While the initial feel of the place was very somber, it softened slightly as I looked more closely. His austere gaze, too, softened, as I looked more closely. One quality though of the initial gist stayed through-out and it governed the experience: The off-center nature of the composition. This of course would be known instantly with the low-resolution part of gist, and this is important. By placing the figure on the left, Velázquez gave him a little personal space, so to speak, that allowed him to be prepossessed, self-confident, and independent. This gives a weight to the figure that it wouldn't normally have if it were centered.

I look to the wall label for the description and was grateful for what language can sometimes give that art and low-level features cannot. (This is why language, while not being isomorphic with art, is often a necessary companion. What we can learn through linguistic communication often dovetails with the visual experience, making the latter much more meaningful.) This was a portrait of Juan de Pareja, who had been enslaved in Velázquez's studio for two decades (a common practice at that time in Spain). A few months after the completion of this portrait, Velázquez legally freed Juan de Pareja, who then himself because a well-known painter in Madrid.

I have given you the process of the perceptual experience, starting with the gist. Please note: What I ended up with was both a reconfirming of my original gist—e.g., the image was off-center, and it was somber in tone—and it also, in some details that proved significant, went partially against that and thus deepened the complexity and profundity of the experience. For example, the color next to his hand was warmer and more inviting than I had originally been able to see. Yet, most of the gist was reconfirmed. For example, the left-hand weight of the canvas stayed the same and continued to drive the effect of a man who was slightly pulled back, slightly unavailable, but also self-confident and prepossessed. The color was the same also from the initial gist as the over-all effect was a still a somber one, which continued to allow me to see the figure as one I should take seriously; there was no over-arching sweetness or demure quality, none that was too soft or available. The change in the color near the hand added a slight addendum to this, and contradicted only enough to add complexity: the color (and the man himself) was now only a little soft and a little available.

This is what looking is. We garner an initial gist level of information, and then further looks confirm or fail to confirm. Most often, the further looks confirm; otherwise, of course, gist would be useless. But in those parts where they fail to confirm are sometimes the ones that, like jack-in-the-boxes, give the most delight. I was happy to find Mr. de Pareja to be more willing to come forward and communicate than I had originally thought; for this man who was, in my gist, leaning away from me and shrouded in somber tones ended up being, upon closer examination, softened and sensitive. It was worth the closer look.

References

Balasubramanian, V., & Sterling, P. (2009). Receptive fields and functional architecture in the retina. *The Journal of Physiology, 587*(12), 2753–2767.

Bar, M., & Aminoff, E. (2003). Cortical analysis of visual context. *Neuron, 38*(2), 347–358.

Block, N. (2014). Rich conscious perception outside focal attention. *Trends in Cognitive Sciences, 18*(9), 445–447.

Burge, T. (2010). *Origins of objectivity*. Oxford University Press.

Castelhano, M. S., & Henderson, J. M. (2008). The influence of color on the perception of scene gist. *Journal of Experimental Psychology: Human Perception and Performance, 34*(3), 660–675.

Crane, T. (1992). *The nonconceptual content of experience*. Cambridge University Press.

Çukur, T., Huth, A. G., Nishimoto, S., & Gallant, J. L. (2016). Functional subdomains within scene-selective cortex: Parahippocampal place area, retrosplenial complex, and occipital place area. *Journal of Neuroscience, 36*(40), 10257–10273.

DiCarlo, J. J., Zoccolan, D., & Rust, N. C. (2012). How does the brain solve visual object recognition? *Neuron, 73*(3), 415–434.

Dretske, F. (2015). *Perception versus conception: The Goldilocks test*. Oxford University Press.

Epstein, R., & Kanwisher, N. (1998). A cortical representation of the local visual environment. *Nature, 392*(6676), 598–601.

Evans, G. (1982). *The varieties of reference*. Oxford University Press.

Fish, W. (2021). *Philosophy of perception: A contemporary introduction*. Routledge.

Greene, M. R., & Oliva, A. (2009a). The briefest of glances: The time course of natural scene understanding. *Psychological Science, 20*(4), 464–472.

Greene, M. R., & Oliva, A. (2009b). Recognition of natural scenes from global properties: Seeing the forest without representing the trees. *Cognitive Psychology, 58*(2), 137–176. https://doi.org/10.1016/j.cogpsych.2008.06.001

Greene, M. R., & Oliva, A. (2010). High-level aftereffects to global scene properties. *Journal of Experimental Psychology: Human Perception and Performance, 36*(6), 1430–1442.

Groen, I. I., Ghebreab, S., Prins, H., Lamme, V. A., & Scholte, H. S. (2013). From image statistics to scene gist: Evoked neural activity reveals transition from low-level natural image structure to scene category. *Journal of Neuroscience, 33*(48), 18814–18824.

Herzog, M. H., Thunell, E., & Ögmen, H. (2016). Putting low-level vision into global context: Why vision cannot be reduced to basic circuits. *Vision Research, 126*, 9–18.

Itti, L., & Koch, C. (2000). A saliency-based search mechanism for overt and covert shifts of visual attention. *Vision Research, 40*(10–12), 1489–1506.

Itti, L., Koch, C., & Niebur, E. (1998). A model of saliency-based visual attention for rapid scene analysis. *IEEE Transactions on Pattern Analysis and Machine Intelligence, 20*(11), 1254–1259.

Julian, J. B., Ryan, J., Hamilton, R. H., & Epstein, R. A. (2016). The occipital place area is causally involved in representing environmental boundaries during navigation. *Current Biology, 26*(8), 1104–1109.

Kanwisher, N. (2001). Neural events and perceptual awareness. *Cognition, 79*(1–2), 89–113.

Koch, C., & Ullman, S. (1987). Shifts in selective visual attention: Towards the underlying neural circuitry. In *Matters of intelligence: Conceptual structures in cognitive neuroscience* (pp. 115–141). Springer.

Land, M. F., & Hayhoe, M. (2001). In what ways do eye movements contribute to everyday activities? *Vision Research, 41*(25–26), 3559–3565.
Land, M., Mennie, N., & Rusted, J. (1999). The roles of vision and eye movements in the control of activities of daily living. *Perception, 28*(11), 1311–1328.
Lim, S., Padmala, S., & Pessoa, L. (2009). Segregating the significant from the mundane on a moment-to-moment basis via direct and indirect amygdala contributions. *Proceedings of the National Academy of Sciences, 106*(39), 16841–16846.
MacEvoy, S. P., & Epstein, R. A. (2011). Constructing scenes from objects in human occipitotemporal cortex. *Nature Neuroscience, 14*(10), 1323–1329.
Maddock, R. J. (1999). The retrosplenial cortex and emotion: New insights from functional neuroimaging of the human brain. *Trends in Neurosciences, 22*(7), 310–316.
Mandelbaum, E. (2018). Seeing and conceptualizing: Modularity and the shallow contents of perception.
Oliva, A. (2005). Gist of the scene. *Neurobiology of Attention 2005*, 251–256.
Oliva, A., & Schyns, P. G. (1997). Coarse blobs or fine edges? Evidence that information diagnosticity changes the perception of complex visual stimuli. *Cognitive Psychology, 34*(1), 72–107.
Oliva, A., & Torralba, A. (2006). Building the gist of a scene: The role of global image features in recognition. *Progress in Brain Research, 155*, 23–36.
Olshausen, B. A., Anderson, C. H., & Van Essen, D. C. (1993). A neurobiological model of visual attention and invariant pattern recognition based on dynamic routing of information. *Journal of Neuroscience, 13*(11), 4700–4719.
Park, S., & Chun, M. M. (2009). Different roles of the parahippocampal place area (PPA) and retrosplenial cortex (RSC) in panoramic scene perception. *NeuroImage, 47*(4), 1747–1756.
Parkhurst, D., Law, K., & Niebur, E. (2002). Modeling the role of salience in the allocation of overt visual attention. *Vision Research, 42*(1), 107–123.
Peacocke, C. (1983). Sense and content: Experience, thought & their relations. *Mind, 94*(375), 480–487.
Peacocke, C. (1992). *A study of concepts*. The MIT Press.
Potter, M. C. (1976). Short-term conceptual memory for pictures. *Journal of Experimental Psychology: Human Learning and Memory, 2*(5), 509–522.
Prinz, J. (2013). Siegel's get rich quick scheme. *Philosophical Studies, 163*(3), 827–835.
Ramkumar, P., Hansen, B. C., Pannasch, S., & Loschky, L. C. (2016). Visual information representation and rapid-scene categorization are simultaneous across cortex: An MEG study. *NeuroImage, 134*, 295–304.
Sekuler, R., Watamaniuk, S. N., & Blake, R. (2002). Perception of visual motion. *Stevens Handbook of Experimental Psychology, 1*(2).
Schyns, P. G., & Oliva, A. (1994). From blobs to boundary edges: Evidence for time-and spatial-scale-dependent scene recognition. *Psychological Science, 5*(4), 195–200.
Schwabe, K., Menzel, C., Mullin, C., Wagemans, J., & Redies, C. (2018). Gist perception of image composition in abstract artworksi. *Perception, 9*(3), 2041669518780797.
Siegel, S. (2006). Which properties are represented in perception? In T. S. Gendler & J. Hawthorne (Eds.), *Perceptual experience* (pp. 481–503). Oxford University Press.
Siegel, S. (2011). *The contents of visual experience*. Oxford University Press.
Tatler, B. W., Baddeley, R. J., & Gilchrist, I. D. (2005). Visual correlates of fixation selection: Effects of scale and time. *Vision Research, 45*(5), 643–659.
Torralba, A., Oliva, A., Castelhano, M. S., & Henderson, J. M. (2006). Contextual guidance of eye movements and attention in real-world scenes: The role of global features in object search. *Psychological Review, 113*(4), 766–786.
Watson, D. M., Andrews, T. J., & Hartley, T. (2017). A data driven approach to understanding the organization of high-level visual cortex. *Scientific Reports, 7*(1), 1–14.
Ballard, D. H., Hayhoe, M. M., Li, F., & Whitehead, S. D. (1992). Hand-eye coordination during sequential tasks. *Philosophical Transactions of the Royal Society of London. Series B: Biological Sciences, 337*(1281), 331–339.

Chapter 2
Seeing: *Sensing*

Abstract Experience begins with sensing, which is significantly of particularities. I call these "p-unitates". How does sensation filter into the process of experiencing artwork? Taking into consideration the last chapter's description of nonconscious perception, gist, and affordances, this chapter lays out the role of sensation in the pull and push of perception. Reality exists already in a schema, and we do not see everything—we edit—and sensing is an important part of gist perception. Thus, the trajectory is not measured in the usual way by starting with the object and attempting to determine how it is that the object conveys predicable information to the subjective experience of the perceiver. Instead, I begin with gist and the sensing that is part of it, and track how an individual's gist can—in consort with others' experience—become constitutive of public objects. Considering that researchers have shown nonconscious perception to exist, and that sensation exists in not just a feed-forward model but a dynamic and feedback one giving us our immediate and affective reaction to a situation, we can establish the important role of low-level features in determining when something *seems* to be a certain way to us. The paradigm example is art.

Keywords Gist · Sensations · Susanna Schellenberg · Gareth Evans · Bill Brewer · Gilbert Ryle · Robert Stalnaker · Low-level features · Particularities · Nelson Goodman

2.1 Introduction

There isn't a satisfactory theoretical model that explains this process of how the data garnered from the initial gist experience causally affects the more extended perceptual experience, particularly within the framework of viewing an artwork. My example at the end of the last chapter was an attempt to reconstruct and give an

D. Shottenkirk, *Art as Cognition: How Gist Reframes the Aesthetic Experience as Conversation*, Synthese Library 521,
https://doi.org/10.1007/978-3-032-00143-6_2

experiential account of the perceptual experience of looking at an artwork in a way that took into consideration the functional role of the initial gist and of the reiterated gist as seen in the saccades. It gave an account of how the experience of standing in front of an artwork is not limited to the perception of semantic objects, while simultaneously giving a real-life narrative account of the actual experiencing of low-level features. But to give a theoretical account of the latter is something different. The difficulty somewhat lies in our conception of that initial gist perception and how we might want to model what is, largely, a sensorial experience. Thus, even more to the point, the difficulty lies in the tradition of what we have to say about sensing itself. Let me try to present the problem, and thereby motivate the search for a solution. Taking into consideration the last chapter's description of nonconscious perception, gist, and affordances, this chapter lays out the role of sensation in the pull and push of perception.

2.2 The Debate in Philosophy

2.2.1 The Problem

Perception is always defined as the act of being aware of something through the senses, and generally it begins at the point where perception is seen as the perceiver's openness to the world. Clearly, the issue of sensing, and how we define it, plays a crucial part in this explanation. In this, the beginning point of perception is usually found in describing objects and their properties e.g., the red apple, and explaining *how* it is that we come to know that that redness is signifying an apple. This seems reasonable: What we want to explain is how it is that the external world e.g., the world of things and matter, crosses the barrier of our bodies and becomes fodder for thought. Therefore, the task is generally accomplished by following the trail, so to speak, of the data as they start in the object and then traverse the sensory interface barrier of the perceiver, trying to track, in that process, the identity of that data. Traditionally, that has been expressed in terms of both the properties and the object itself.

I, of course, think that beginning with the presupposition that it is *solely* objects and their properties that we are investigating is a mistaken presupposition. I will also show that to start with the object is a mistake as well, as we are better advised to start with gist and to then track that phenomenal experience in its path of developing linguistic naming. (See below: Sect. 2.2.3.) Global scene perception is at least as important, if not more so. But in the next few paragraphs I will proceed with the discussion in the more traditional way as I am conveying both the problem as it has been traditionally viewed and explaining how that view fails to solve the difficulties.

In that transmission from object to perceiver, philosophers often ask whether those sensorial experiences are still of the external, material object, or are they now sensorial, subjective, or even mental? What really is it to sense *red* qua red? And how exactly is that connected to the perception of the object apple qua apple? We might want to think of properties as something roughly analogous to low-level features: shapes, colors, textures, size, luminosity, etc., and then objects would fall neatly into the category of high-level/semantic features e.g., the ball, the tree, etc. I want to caution against this, at least at this point, for reasons that are primarily about the metaphysical pitfalls that might accompany such a move: Properties are often thought of in platonist terms, and that is not the direction I wish to go, nor is it how low-level is normally thought of. Instead, I would like to point out that the question (e.g., regarding the divide between those who think sensing gives us direct experience of the object itself and those who, on the other hand, argue that the experience is of the sensing process itself), is a divide that tracks those who advocate for direct perception or even a modified notion of direct perception e.g., Bill Brewer's relational account of perception (Brewer, 2007, 87–97) or those who, like Humean phenomenalists, argue (directly) for indirect perception. Whether or not one wants to take the side of the direct perception or indirect perception theorist, they often share the following presupposition: That both object and subject are completely discreet and separate entities, a problem magnified if one also wants to argue that they are of different ontological kinds.[1] For we therefore have a conundrum when we try to explain their interaction: When you have a complete bifurcation between the constituted natures of the thing perceived and the perceiver, it becomes enormously difficult to explain how the properties of the external object become part of the experience of the perceiver. In other words, *How exactly* does the external world cross the barrier of the sensing perceiver and become fodder for thought?

What underwrites this difficulty is the fact that the traditional analysis of sensing used something like a mechanical model dependent on a feedforward process: First the data arrives at the doorstep of the perceiver, and then the sensing mechanism transduces that data into useful information allowing thought to happen. But the difficulties here, too, are enormous. For, once we have the faculty of perception in charge of absorbing the sensorial data from the world and the—quite different—faculty of cognition as operating on that bottom-up data, we must then decide several things: (1) which to prioritize (e.g., is matter of more importance or are mental ideas), (2) how to determine when one stops and the other begins, (i.e., a problem suffered by most vision scientists today, and many philosophers), and (3) how to account for the fact that there is this mysterious transduction that happens when the world "out there" becomes part of the world "in here" (i.e., the problem of "the

[1] For the traditional idealist, for example, humans would have abstract objects as constituents of their mental faculties, while physical objects would of course not; reductivist materialists, while viewing both humans and inanimate objects as material, might though also see a great divide between humans as subjects and therefore metaphysically distinct from objects. In any case, there is the tendency across many different philosophical positions to regard human subjects as being on one side of the existential chasm while objects are on the other. This radical bifurcation creates substantial problems for epistemology.

given" suffered by the Logical Positivists in trying to answer this within an empiricist framework).

All of these problems have been tripping stones. I argue that some of these issues can be put to rest if we, instead of seeing this as a two-pair relata (i.e., perceiving and cognizing), and viewing this in quasi-mechanical terms, we instead update our framework and see it in terms of a dynamic process that involves both feed-forward and feed-back loops. We also need to give an account of how our editing of the plethora of data in gist interacts with our process of naming, and why it is that sensing is so important to us. Thus, we can see some way to give a more robust and accurate account of perception. And of cognition. Or, more properly, an account that succeeds in no longer dividing the two to the point of incomprehension.[2] I hope to go some distance in resolving this problem and believe it can be done once we start with gist experience.

2.2.2 *Seemings, Particularities, and P-Unitates*

What exactly is sensing? It is Thomas Nagel's oft quoted "What it's like" that points us to the irreducibility of the sensing experience, and most cogently brings to mind the importance of sensing. In his arguments against a reductivist move, Nagel argues that consciousness is simply the experience of the subjectivity found largely in consciousness consequent to sensing, and that that cannot be explained in any other terms—third-person descriptions do not even get close to describing first-person experiences. Thus, in his famous example, we cannot, in theory or in fact, know what being a bat is like. To know the "feel" of any experience is ineluctably private, and yet immeasurably valuable despite its resistance to quantification.

Various contemporary philosophers have written on sensing, notably Christopher Peacocke (Peacocke, 1992), but it is Gareth Evans' comments that are most pertinent for our purposes here. This was briefly referred to in Sect. 1.2.1, but it will be explained more fully herein. Writing in the 1980s, Evans' posthumous *The Varieties of Reference* addressed concerns of representation and semantics, and significantly emphasized the role of sensing, most poignantly in an oft quoted (including earlier in this text!) footnote: "…the senses yield non-conceptual information, whereas language embodies conceptual information…(Evans, 1991, 123 fn 5)". For Evans, objects are not only perceived in terms of being an instance of a named category e.g., "apple", but instead are that part of the information that we receive from perceptual experience that comes from sensing, and *that* is non-semantic.

His emphasis on sensing was not new, of course: One need only think of Russell's distinction between knowledge by description and knowledge by acquaintance. Evans' noted this issue in Russell and wrote, "All the items with which a subject

[2] See Andy Clark's research on predictive processing for another way out of this problem (Clark, 2015).

may be acquainted have what may be termed the Cartesian property…But Russell, like any Cartesian, would wish to decompose such predicates into strictly mental and non-mental components, the claim about infallibility extending only to the former" (ibid., 44–45). In many ways, Evans is not opposed to Russell and is trying to explain perception such that it doesn't run afoul of what he calls "Russell's Principle" e.g., a subject cannot make a judgement about something unless he knows which object his judgement is about (ibid., 89), while yet explaining, contrary to Russell's Cartesian requirement, that we make errors in perceptual judgements. This is an interesting point that Evans is articulating—i.e., that we make errors in perceptual experience. And though I disagree with some of the points of emphasis that Evans makes, (such as over-emphasizing the fallibility of sensing while largely ignoring the fallibility at the specific locale of our inductive generalizations to semantic categories), I largely agree with the direction in which he is going. As Evans states, "However, there does not seem to me to be anything incoherent in the idea that it may be, for a subject, exactly as though he were thinking about a physical object (say) which he can see, and yet that, precisely because there is no physical object he is seeing, he may fail to have a thought of the kind he supposes himself to have" (ibid., 45). We can, in his example, mistake two moving balls for one. Because it is obvious that perception is at heart an imprecise venture, it is therefore important to note the ways we make mistakes, and thus Evans was correct to attempt to describe how our perceptions error, for it is only then that we can form a theory explaining successful perception.

The sort of information garnered with sensing was what Evans called "*seemings*", a descriptive term for how things seem to be at the initial sensorial stage. Evans is pointing to the fact that while something might, initially, *seem* to be a certain way to us, we can yet be mistaken about that initial sensing information—that is, our later sensing can diverge. We look, re-look, we edit, we reevaluate. This is why it is important to model perception on feedback loop processes, and not on a quasi-mechanical feedforward view, for feedback loops can instigate corrective edits and force divergent aggregate generalizations.

Let me say a little bit more about what I agree with here and about what I disagree. Evans wanted to put seemings in a provisional category: he was aware that our sensing experience is rapid-fire and that we check, re-check, edit, re-edit, and can make mistakes. That is true. But, although he didn't directly claim that the linguistically defined categories of semantic objects provided a measure of certainty[3] or at least less uncertainty than the seemings, it might inadvertently seem at first blush that it was only the category of seemings that was dubitable. But that would not be true. As Robert Stalnaker wrote:

> Evans regards it as important to identify information-bearing states of perceptual systems with states of *seeming* since he is anxious to avoid the traditional epistemologist's picture

[3] For further reference see his discussion of the generality constraint: "if a subject can be credited with the thought that a is F, then he must have the conceptual resources for entertaining the thought that a is G, for every property of being G of which he has a conception (Evans, 1991, 104). This is discussed in Sect. 4.3.2.

> according to which the subject receives, through the perceptual systems, sensory data that is 'intrinsically without objective content,' but which forms the basis for inferences about the world that causes them (Stalnaker, 1998, 352).[4]

Stalnaker is right: Evans' main target was the epistemological point of view that sees information as coming to us in an objective manner i.e., as a pure "given", without being part of a pre-existing schema. That point of view—the "traditional epistemologist's picture"—was that of the British Empiricists; but, as Evans was well aware, the failure of the Logical Positivists' program showed us that the notion of the pure and objective given had been discarded. Instead, all information comes to us already placed within a context and a schema. The world comes to us already embedded in systems of meanings. This applies to semantic objects just as much as to the particularities of sensed seemings.

But to say that information comes to us in an already given system where the meaning is at least somewhat dependent on context, as well as the prior systems of experience stored in memory, is not say anything about whether the sensed entity is necessarily externalist. In other words, this discussion has so far been neutral on whether or not sensing is purely phenomenal or whether talking about sensing commits us to veridical perception. If semantic objects are already embedded in systems of meanings antecedent to our perception of them, then not only does context play a role but so does both pertinent cultural norms as well as the perceiver's personal history. Thus, the question of objectivity comes into play.

The debates surrounding the "common factor"—i.e., those shared by veridical perception, hallucination, and illusion—have been voluminous, and while my interest in not there, I will point out the worthwhile contributions Susanna Schellenberg has made, particularly emphasizing her argument for a moderate externalism, where the content of the perception is only in part constituted by the subject's subjective experience. She states, "We can distinguish the uncontroversial idea that perceptual experience manifests phenomenological particularity from the controversial idea that perception instantiates *relational particularity*. A mental state instantiates relational particularity if and only if the mental state is constituted by the particular perceived" (Schellenberg, 2016, 28).

"Particularity" is the term Schellenberg uses to draw our attention to the act of perception itself: We perceive particulars. "Perceptual experiences are (as) of particulars" (ibid., 27). While this point involves too her broader notion of capacitism, where to perceive is to enjoy personal level perceptual capacities, I will leave aside the broader point of capacitism and discuss instead her notion of particularities as that relates more closely to the topic of this chapter. She makes the point that the perception of particularities "…is a claim about the sensory character of the subject's perceptual state, not the conceptual content of any judgment she might form on the basis of her experience" (ibid., 28). In other words, we have a phenomenal

[4] The "traditional" epistemologist here would include the (Humean) empiricist as well as both the radical empiricist who views objects as the neutral given and the realist who views our reception of external reality as direct, completely mind-independent, and thus "objective".

experience which is separate (in principle) from our ability to frame that experience as an instance of a larger conceptual or linguistic category as well as separate from our ability to confirm the veridicality of that experience. While not arguing for disjunctivism, (i.e., where there is no common factor between veridical perceptions and hallucinations/illusions), Schellenberg still wants to recognize some substantive common element. But she is nonetheless focused on providing a grounding for naturalism and a modified externalist epistemology. She states, "A mental state manifests phenomenological particularity if and only if it seems to the subject that there is a particular present" (ibid.).

For Schellenberg, the goal of perception, by providing information about particularities in the environment, is largely to fix the reference of singular terms. But reference is not the ultimate criterion. Unlike a philosopher like Nelson Goodman, who grounded everything in reference, Schellenberg recognizes that perception is much richer and more complex than that:

> …we can say that discriminating and singling out a particular from its surround is a minimal condition on perceiving the particular… Non-rational animals and infants as young as four months old can perceptually single out objects and property-instances in their environment, yet they do not have the capacity to refer. While referring requires conceptual capacities, singling out particulars requires no such capacities (ibid., 36).

This is an important point. And while she will want to tie relational perception (and its veridicality) to semantic content, it is nonetheless important to spend some time on the fact that perception itself is not so limited. When we perceive the particularities of our environment, some of this is concatenated into the ontological chunks of proprietary objects with their linguistic referents, and some is not.

This pulls us back again into the discussion of low-level features of perception. The fact that the perceptual process is grounded in the singling out of particularities shows that the process is tethered to the bodily processes involved in not only foveating and saccadic scanning, but the relation of those retinal processes with the many parts of the brain that coordinate in order to experience perception and for recognition to happen. Thus, perceiving is a bodily process.

It is this that is worth our attention. Unlike Schellenberg, I am not particularly interested in distinguishing the process of perceiving phenomenal particularities from the process of perceiving relational particularities. For that distinction is one that tracks the role of semantic content, and I am interested not in language but instead in art. And art relies very significantly on the perceiver's (as well as the artist's) phenomenal experience of low-level features, which themselves often escape the constraints of language. The roles of truth and of an externalist account of perception are thus different in art than they are in language.

Thus, sensory particularities, in my view, are not limited to either objects or to semantic content. They are not properties or predicates (as these are defined in terms of their relations to objects), and they are not limited to conscious perceptions. And they are not defined by the notion of particulars as opposed to universals. Instead, the particularities that I am identifying are the basic elements—what I will call the "p-unitates"—of all perception, including that of low-level features. These are, in turn (and in some instances), derivatively the constituents of high-level features, and include both veridical and non-veridical perception, as well as conscious and unconscious perceptions. I use the term "p-unitates" as the term allows us to see the commonality between phenomenal particularities and relational particularities. Whether the p-unitates are only phenomenal subjective experiences or whether they are the experiences derived veridically from the physical, external world (or ones that *could be* derived veridically from the physical, external world) can be left underdetermined. Thus, these p-unitates are the determinables of experience even in those cases when are not yet determinate.[5]

Thus, there is a relationship between the constituents of subjective, unsubstantiated, phenomenal experiences that turn out to be veridical and those phenomenal experiences that do not e.g., hallucinations/illusions. The perception involved in both veridical and non-veridical experience thus has to be thought of as a temporal process, where the initial phenomenal particularity has a character that is similar in both cases and it is only the *subsequent, later maneuvers*—the fact-checking, the confirmations rendered by others' opinions and evidence—that moves those phenomenal particularities from the category of unsubstantiated to substantiated. Or, alternately, from the category of believed-to-be-substantiated to the category of the unsubstantiated. This is the role of evidence in belief-formation.

But the important point is to see the similarity of these p-unitates at the beginning: my sensory experience of the particularities in an illusion (or hallucination) will—before the process of fact-checking and the confirmations rendered by others—be indistinguishable to me from the sensory experience of the particularities in a veridical perception. That of course is because research has shown that the heavy lifting, so to speak, of perception is done by the visual cortex, with a sparsity of data provided by the retina to the lateral geniculate nuclei (LGN) limiting the initial role of the externalist world (Chariker et al., 2016). Because the beginning steps of the data acquisition is always very minimal, the sensory quality of the gist strikes a similar chord whether it is an illusion, a hallucination, or a veridical experience. What something "seems" to us at the first blush can be confirmed or disconfirmed. Thus, since we begin with very little in perception, those beginnings have similar character.

But our brains have figured out how to—at least partially and sometimes imperfectly—compensate for this sparsity of perceptual data that flows through the retina to the LGN and then to the visual cortex. We can match the incoming retinal data—however sparse it might be—with past experiences and that is the richness of the visual cortex as it operates in conjunction with the limbic system and the many interconnected other parts of the brain that are in the causal relations with those systems. We see and then say "ah, it is an apple". And, further, "this apple is an example of the general category 'apple'". And: "I just love apples! *So* glad I saw that one!"

[5] This insight I owe to a conversation with Sascha Benjamin Fink of the Otto von Guericke Universität.

Our quite different ability to generalize and think about larger abstract categories and general terms gives us the ability to maneuver in the world—this is true: To take short cuts, to predict, to survive. But it is in the experience of the fine granularity of particularities, including importantly low-level features, that gives us the "what it's like" feature of life. What we are able to pick out from the vast plethora of data as it comes to us allows us to not only survive, but to enjoy the world. The particularities in perception not only are what give us the "what it's like" feel of phenomenal experience, but it also yields the larger experience of finding meaning, which is crucial in this traveling forth in life. And it is the richness of low-level features that significantly constitute that experience.

Let me put the point more directly: It is in the particularities of experience where life lives. It is in the moment of "stopping and smelling the flowers", to use a hackneyed expression, that gives us a connection to the moment's particularities; it is where we find ourselves no longer in the anxieties of the future or the regrets of the past, but in the sensorial moment. As a small aside, it is worth noting how humans have developed various means to make central that experience, with meditation (and medication…) being a central one. Meditation is framed to put us into that kind of awareness of being in the here and now, appreciative of the moment. Whether one practices traditional Mahayana meditation, which focuses on emptying the mind, or Theravada, which instead practices, for example, the mindfulness of focusing on things around one, Buddhist meditation has instilled in its practitioners the discipline of stepping away from the maelstrom of worries, planning, and problem-solving that so plagues the human brain, and yielding instead to the calming pleasures of present particularities. These same practices are seen in other religions, as well. This is the reason that many meditators describe the immediate sensation after meditating as not only calm but having a heightened awareness of the atomistic fracturing of the objects in the environment; a world not broken up into useful and useless objects, but a democratization of perceptual worth—all glittering, all luminescent. The world of sensing.

2.2.3 *Editing and Naming*

To reiterate: Perception is not passive and the world does not simply come to us; I match, I edit. Thus, I too follow in the lines of those who argue that it is not the case that all of reality comes to us unbidden and on its own terms. Our faculties do not just mirror reality.[6] A known entity sits next to many others that are not so well-known or not known at all. There is selection, and editing, and re-editing. It is only in theory that *all* entities *could be* noticed, they *could be* named and categorized into larger semantic categories. But, in fact, that is not the way it works: For to have all entities noticed and attended to would be to have *no* entities noticed and attended to

[6] See Nelson Goodman for discussions of this (Goodman, 1976, 8). See also (Shottenkirk, 2009).

as it would do away with saliency maps and other selection mechanisms that we must necessarily have in place in order to perceive. Thus, choices must be made, making perception something like a relative term—to perceive something e.g., to pay attention to something, means that other entities were *not* paid attention to, were *not* perceived. And so it is: Some things—in fact, most things—do not get noticed.

In order to give further support to this, I shall first reiterate and summarize some of the issues dealt with in this and the last chapter, and then present some of the vision science research behind the issue of perceptual editing. To reiterate. There are two main categories that influence choice. Firstly, there are the biological determinants in the saliency maps e.g., issues of contrast, orientation, etc., that operate on low-level features in a purely bottom-up manner. Secondly, editing happens as a part of the inductive patterns of noticing that are subsequent to past noticings: Once *some*thing has been pointed out and/or attended to, those particularities and/or semantic categories get noticed, making them the constituents of the individual's phenomenal experience.[7] That experience in turn has consequences, for now it provides further predictive power as the perceptual experience gives us the ability to, in the future, *expect* the data and *recognize* it when we see it. (The last subsection of this chapter will elaborate on this issue of editing, bringing in the related notion of naming.)

Again, this issue is made clear in gist experience. For example, we recognize the faces of famous people as we flip through the selections of movies online. Part of what happens in this instance of gist experience is due to the fact that there are sufficient contrasts, differences in line orientation, etc. These are the selection mechanisms that drive many of the bottom-up features in the saliency maps. But there is, as well, the important matching functions that happen when the sparsity of retinal data is transferred from the eye to the subcortical LGN and then to the visual cortex, which in conjunction with other parts of the limbic system and the implicit and explicit memory systems, allows us to know that we have seen these faces before; they are part of our past phenomenological experiences and now part of our inductive p-unitates patterns. We know those faces and have come to expect them in the world, hence *we see* that data. And we can do that in 200 ms (Oliva & Torralba, 2006).[8] And the process is not entirely conscious; often we act on information without being fully conscious of our perceptual recognitions. I might not consciously say, "Oh, that actor is John Smith and I don't like him", but I will—in 200 ms—see John Smith and flip on by him.

The main point that I'm emphasizing here in this subsection is that we more readily notice things for which we already have had prior experience, or for which we have a name, or about which we care. For example, and regarding the latter, the psychology researchers Gantman and Bavel have demonstrated that we edit on the

[7] This may be not quite a complete statement: Nonconscious perception might also be included in the phenomenal experience, but future research will establish if this is true or not.

[8] There are those instances of phenomenal awareness that are exogenous non-predicative cases, for example, a car's backfiring or a slap in the face. This is non-edited information that suddenly intrudes. But I am not interested in those as they are not especially relevant to the gist experience.

basis of moral beliefs: "…that personal beliefs, moral identities, or moral motives may influence the basic awareness and interpretation of moral stimuli prior to action" (Gantman & Van Bavel, 2014, 22). In this, the researchers tested whether we see moral words more than non-moral ones by testing for the moral "pop-out" effect of stimuli that was presented at the threshold of perception e.g., at gist. Building on the knowledge that we see things we care about such as seeing food words when we are hungry more than when we are satiated (ibid.), the researchers were able to extend that principle to percepts that excited moral responses. To give a different sort of example, you could imagine a person who would see fur coats on the street more readily after having had joined PETA; we see what we care about it. The researchers were able to confirm that we did, indeed, see things that re-confirm our moral commitments. Editing is a central part of our perceptual process and it happens at the very beginning of perception.

The physiological underpinnings of this are interesting, as they involve the joint workings of different parts of the brain, particularly of the amygdala and the visual cortex. The amygdala was originally thought to be only the center for fear, but recent research has shown it be involved in most parts of the brain's activities, including the functions of memory, reward circuits, limbic, and cognition, and including as well many emotions other than fear (Avecillas-Chasin et al., 2023). As the promiscuous kids on the block, the two amygdalae connect to more other parts of the brain than do most anatomical structures. As has been noted previously in the text, it is the parahippocampal gyrus that is responsive to global scenes including houses and buildings, while the fusiform gyrus is responsive to faces. Researchers, using this division, tested for how subjects are able to separate unimportant and mundane data from significant data, and what parts of the brain were doing so, by establishing a learning phases where the subjects were shown images of houses (CS+) or buildings (CS-), with the former associated with a shock at 50% of the time (Lim et al., 2009). This of course is testing for fear conditioning (one of the main jobs of the amygdala), and thus is testing for a prime way of separating out the mundane from the significant. As in many of these sorts of visual tests, these researchers used rapid serial visual presentation stream (RSVP), where visual stimuli are presented rapidly (from 6–10 images/s) at the same locale/space. It is the main way researchers have for testing for attention across time, and, in general, when this is done the subject's detection of the second image (T2) is degraded due to the competition for attention demanded by the first image (T1). In this experiment Lim et al. were testing for how fear would increase the percentage of instances when the subject was able to perceive T2, thus testing how changes in the amygdala translated into changes in the visual cortex as well as behavioral responses. They write, "Our main findings can be summarized as follows. Behaviorally, affectively significant scenes (CS+) were better detected than neutral (CS-) ones" (ibid., 16845). In other words, when the subjects' amygdala registered something of importance e.g., "we need to be afraid of CS+", the visual cortex was better able to see it. Editing happened.[9]

Naming thus has an important role in determining what we see and what we don't see. And it does it in two ways. Firstly, in a more trivial sense perhaps, naming partially controls *what* we see. We have touched on this already, so suffice it to quickly summarize. Naming allows us to see things that would, if we had not had names for them, likely go unseen. Research has shown not only that we are prone to seeing meaningful images in vague and nebulous visual stimuli such as a face in the stains on the sidewalk, or a human figure in the tree bark (a phenomenon called "pareidolia"), but we are more easily able to identify objects for which we have a name. For example, if you imagine not knowing what a computer server looked like but you came upon one for the first time, all the rows of shelves and lights and boxes with handles would not ontologically congeal as one entity, and instead, it—the server qua server—would not be seen. You would simply be seeing a grocery-isle of identical boxes with handles. In this rather trivial way, we see what we antecedently know. The same is of course true in social phenomenon: If a person had never been taught what racism or sexism was, for example, that person would simply fail to see many of those instances. Knowing, in some measure, comes before seeing, and not after.

Secondly, and in many ways much more importantly than the issue of how naming partially determines what we see and don't see, there is the question of the general causal connection between sensing and linguistic naming: How do we get from one to the other?

In regard to this more difficult issue of how we get from sensing to naming, let me for a moment return to the discussion in Sect. 2.2.1 and explore some of the difficulties faced when we assume the traditional separation of perception and cognition and see that process in a quasi-mechanical way. If we were to agree (which we should not!) that it is this two-step process of (1) perception of raw sensory information and then (2) cognitively processing it into semantic categories, then it is difficult to explain *how one learns* to identify semantic objects on the basis of seemingly unrelated sensory data. The relationship between the non-cognitive percepts and the high-level cognitive, semantic categories must start somewhere.

There's a history of trying to account for this in philosophy. For example, Gilbert Ryle made this criticism in the following statement:

> We inadvertently work on such models as that of a solitary man inside his tent who sees spots and patches of light and feels indentations in the inside of the canvas. He then, perhaps, wishes he could see and feel the torches and boots that made those patches of light

[9] The authors of this study though caution against interpreting this relationship between the amygdala and the visual cortex in a determinate causal relationship as there may be a more indirect explanation: "It has been suggested that frontoparietal regions are involved in the control of visual competition, including competitive interactions during the AB, and that they may be critical 'bottleneck' sites…For instance, the role of frontoparietal regions may be better viewed as one of 'gating' the link between the amygdala and visual cortex, namely, a more moderation-like interaction pattern" (Lim et al., 2009, 16,845).

> and indentations in the canvas. But, alas, he can never see those torches nor feel those boots, as the canvas is always in the way (Ryle, 2009, 209).

Let me say a little about Ryle's notion of the term "sensation" and our current use of the term "sensing". For Ryle, "sensation" was a problematic term, and only applied to tactile or kinesthetic experiences and adamantly not to seeing, hearing, tasting and smelling. The latter modalities involved observation, whereas the former did not. His complaint about this confusion regarding the terminology was directed at both (1) the sense datum theorists who posited sense datum as independent entities, (whether mental or non-mental), which were underwritten by platonist structures, and (2) the subjectivist theories of indirect perception held by such people as Hume. In both of these oppositions, Ryle was adamantly opposed to such transcendent entities, and wanted language grounded more firmly in the day-to-day embodied experience expressed as live-language sentences.

I would dispute some of this in Ryle but not all. For example, I don't think his firm distinction between the modalities of seeing, etc., and those of tactile and kinesthetic experiences are completely valid, as I see some area of overlap between the purely "inner"/internalist, so to speak, experience of touch and, on the other hand, the modalities of seeing, hearing, etc., which can more easily be given an externalist account. It seems to me that the more proper way to see all of these modalities is on a spectrum, and not as a bifurcation. Touching something (or being touched) involves, too, some externalist notion of being causally determined by the exterior environment, although one's awareness of the impact on one's body is more acute than it is in those situations as in seeing (where light has hit one's retina), or in hearing something (when sound has hit one's inner ear). The simple fact is that we do not have nerve endings on the ends of our retina or inner ear that transmit to us our awareness of the moment of impact. Skin is different in that way. So, when skin is pressed our nerve endings transmit to us a sensation that our retina and our inner ear do not. But that difference is not acute enough to, in my view, warrant Ryle's bifurcation. Thus, I would argue that the modalities of seeing, hearing, tasting, are different in degree but not in kind from the modalities of tactile sensations. The former *do* give us a more acute awareness of our relation to the thing that is the cause of our organismic response as an external thing of which we are aware and of which we are observing. But there is also an awareness of the causal role of an external thing when we are touched or are touching something, although in a somewhat diminished role. This is true even with pains. We are aware of a causal agent. The difference is that sometimes that causal agent is inside our own body, thus muddying the externalist account in practice although not in theory.

But even having said that I think Ryle was wrong in his (too strident) separation and I would like to turn him on his head in regard to that, I do though agree with Ryle's attempt to steer philosophy away from the rationalist tendencies that place meaning in platonic realms, whether those be purely idealistic or the abstract objects of propositions that in turn govern content. Instead, it is lived sentences that convey meaning and they do so as context-driven, embodied experiences.

Other philosophers too, have been sensitive to the general problem of how we learn to develop symbolic and abstract relations, and I would offer Locke as an example of this. He gave a model of childhood learning regarding the specific object, but Locke's was an explanation of *how we come to general terms*, and in this does not fully answer Ryle's complaint. Let me present the problem in some detail. For Locke, an individual experiences and identifies objects on the basis of matching those experiences with prior ones. For example, this thing before me is an apple and I know that because I've seen apples before. But this iterative process must have a beginning. All the sensing of red, round, size of my hand, not-too-hard, not-too-soft, smooth, etc., doesn't *necessarily* trip my brain over into the category of "apple". Instead, I must have learned that in the past. But how to explain that first instance? How did the first cognitive experience get its foothold? The usual explanation involves the notion of naming, and this was the go-to explanation for Locke as well. Parents or caregivers tell the child "apple", and this word becomes associated with the object (Uzgalis, 2007). Again, Locke dealt with this part of the process, and that seems satisfactory on the surface. But there is a bigger story here that gets left out. What was the starting point? How did "apple" ever get attached to apple?[10]

We could give an anthropological story, but it wouldn't be very interesting. There is a more relevant point, and it is this. There are two different kinds of "starting points" and I would refer to Aristotle's articulation of this in the *Posterior Analytics*: There is a difference between "what is prior and better known in the order of being and what is prior and better known to man (Ross & Smith, 1910, Book I 72a)". Locke's explanation falls into the latter category, whereas what Ryle was looking for falls into the former. If we think about the order of events from our own solipsistic point of view e.g., the order or explanation that is "prior and better known to man"—we think of the temporal order of events in terms of the sequencing of them as those events have happened to us. So, we would analyze how names come to be known by seeing the timeline of events in a child's life and articulating that sequence as they impact that child. This was Locke's version of how we learn names.

But, if on the other hand, we want to know, so to speak, the bird's eye view of what must *necessarily be in place* in order for the events to exist, then we have "what is better known in the order of being". This was Ryle's worry: What must exist in order for language and naming to be understood? The answer for a worry like Ryle's i.e., how can the symbolic words get attached to sense datum if all we start from is the sense datum?—can be easily transposed into present purposes in regard to sensing. For Ryle, they could not, which was a reason to discard sense datum theories and a reason to not call seeing a kind of sensation. But, if we instead view modalities on a spectrum, and think of them as sub-species of sensing, the problem is here as well: How do we take the phenomenological experience of sensing—which is embedded in the phenomenological experience of gist—and convert that into words and into affordances?

[10] See or an interesting account of this see Nietzsche's discussion of how we falsely believe words encapsulate things (Nietzsche, 2005, 14–25).

The answer is largely found in a social epistemological account of language and perception. As it was to Ryle, it is clear to us that we cannot, with one fell swoop, go from sensing to language. We must instead take account of the fact that the existence of a named object assumes a substratum of past sensorial experiences by individuals[11] who, through their cumulative and shared articulations of those experiences have acted together to name that object. So, that object only exists *as an object* as *the consequence* of multiple and shared experiences. This is the ontological consequences of shared sensing experiences: *We* construct objects.

Thus, those experiences of multiple perceivers then accrue together in the form of shared linguistic practices, which are multi-factorial and can apply to various constituent parts of the object via the identification of general terms that bound their set of instances. For example, we identify the general term "apple" as having, potentially, under its umbrella individual instances that instantiate such predicates as varied as red, yellow, rotten, delicious, etc. That general term has come to exist as a result of the various linguistic practices that have identified the individual instances of "apple", and *those* individual instances exist *as a consequence* of being the source of the sensorial experiences of an individual perceiver.

This is of course what underwrites Locke's account; we know names on the basis of others' perceptions. Ryle's concern gives us something different. And that is this: Ryle's notion of observation, which was at the heart of perceptions involved in seeing, hearing, etc., bring us to the external environment. This saves us not only from internalism, thereby giving us a foothold in an externalist account of epistemology, but also—in a surprising way—allows us to theorize independently of the traditional perceiver/object dualism.

This is an important point. Let me take it further. Experience begins with sensing, and in this, it is logical to shift from an account that prioritizes objects to one that starts with global scenes. For sensing is significantly of particularities; this was Nagel's point. Those are the things that are impossible to translate into a third-person account, and yet through their multiplicity and interwoven complexity give us a sense not only of the basic constructs of the external world but important information regarding our relation to that world. The low-level features of disorder/order, vague shape, basic color, line orientation, contrast, etc., etc., are perceived in milliseconds for very good reasons. We need it immediately and use it to go forward. And like with objects, the perception of the global scene does not come without context. The low-level features of the experience, while significantly operating in conjunction with biologically determined saliency maps, also feed into the larger (inductive) semantic categories that are, in turn, also perceived in gist e.g., indoor (the generalized shape of buildings) versus outdoor (the generalized shape of spatial openness), orderly versus disorderly, etc. We have learned to identify these global semantic categories, as well as the constituent low-level features, on the basis of not only our own individual past experience, but the past experiences of others that are

[11] For an interesting discussion of this and related problems see Bill Brewer's "Consciousness and Content in Perception" (Brewer, 2017, 41–54).

passed down to us in the form of naming. For example, "this is an outdoor scene". We learned that—*those words*—from others. And then applied it to our gist experience. In other words, the individual's sensorial experience is completely underwritten *by the prior* existence of the socially shared existence of that same thing, as well as the social shared commitment to the name.

And here is the main point: To do an epistemological analysis of how-we-know-the-world with the construct that one starts from the antecedently existing object and then traces how that object becomes an instance of experience to the perceiver, is surely to misconceive the relationship. It is not the case that perceptual experience gives us the raw data in non-cognitive form that then, afterwards, our cognition operates on forming thoughts in that domain. That is to adopt a too static and over-simplified model when in fact a dynamic system model is needed.

There are further complications. Perceptual data exists in a temporal and provisional order. This means that not only is data dependent on previous data, but also on later perceptions. This may seem like the cart before the horse, but if one thinks of how neural nets work in computed learning, it is then obvious that the accrual of emphasis (e.g., the building up of the nodes in the neural net model) increase or diminish the importance of a particular piece of information, much in the same way that re-iterated memories can increase the importance of a past event. Induction is not just a feed-forward model but a dynamic one that is constantly remaking the past in the same moment as it is predicting the future. That is how we think.

Thus, perceiving is not a predetermined or a priori path. In any moment of experience, we are re-emphasizing certain items because that is what is needed at that moment, which in turn has effects both going forward (e.g., we will more likely perceive those things in the future) and also has an effect going backwards (e.g., we will connect the present with its antecedents in that past, thereby emphasizing those antecedents). In addition, of course, there are many things that are missed. Not all is seen and recognized.

The problem as I see it is the point at which the analysis begins. If we begin, as do those theorists who identify perception and cognition as mutually exclusive, *with the object*, then we are stuck with trying to figure out how the act of perception i.e., *of the object*, becomes fodder for thought, a question which then leads us to wonder how the object comes to reside in our mental faculties. This is to ask for trouble. To begin with the object seems like a dead-end.

But to only switch to global scenes is an answer albeit a partial answer only. What is needed in addition is the following: *tracking the experience as it becomes a public object.* Or to phrase it slightly differently, the trajectory I see is not measured in the usual way by starting with the object—ascertained by perception and then operated on by the independent faculty of cognition—and attempting to determine how it is that the object conveys predicable information to the subjective experience of the perceiver. Instead, I begin *with the gist and track how an individual's gist can—**in consort with others' experience**—become constitutive of public objects.* The answer, which will be expounded on in the rest of this book, is the following. Gist is what we use to make objects—to ontologically chunk the world—which are constructed in consort with our affordances and how we measure our relation to the scene and its objects. The paradigm example of this is art.

But to understand how this works means that we have to understand the precise way a perceiver sees, how an artist makes, and how the two converse. The rest of this chapter lays out the ways we gain information through sensing, with the next chapter summing up the role of the perceiver by laying out a new way of looking at perception; we then proceed into an analysis of how artists perceive and how from that perceptual experience they make art, followed by an explication of how communication between the artist's artwork and the perceiver is a kind of conversation, a sort of social epistemology. At the end of this, we will have a grounded idea of the role of gist in not only our perceptual faculties but our creative ones as well, and gain from that a picture of the epistemic functions of art: Art as conversation; art as cognition.

2.3 The Rise of the Nonconscious

2.3.1 In Psychology and Philosophy

We often want to think of thoughts as necessarily conscious, and yet the last 150 years or so of psychoanalysis has convinced us otherwise. Whatever complaints one might have of Freud and his misguided notion that women's sexuality revolved around penis envy and necessary feelings of inferiority, he did succeed in making a significant dent in all of humanity's belief in our transparent self-knowledge. Repression is the rule: Much is buried, much is lied about, and we too often remain a mystery to ourselves. Forget the philosophical problem of "other minds"—it is the mystery of our own that befuddles us the most.

Beliefs too might be a thing claimed as something that is only held consciously, yet psychology has revealed the extent of nonconscious bias in our daily lives and has thus dethroned the notion, too, that beliefs are conscious and completely under our control (Noon, 2018). With many studies revealing the top-down control of nonconscious beliefs upon our actions and attitudes, including those with racist and sexist content, we have all become justifiably wary of what cruel thoughts lurk beneath our daily habits.

A similar dethroning of conscious beliefs has taken place in philosophy. To give one example: Tim Crane positions belief as primarily unconscious, as it is the sort of thing that subsists even when we are asleep or unconscious, thus making beliefs instead the paradigmatic example of unconscious mental states (Crane, 2013).[12]

Alas, so much for thoughts and beliefs. Let us then consider now the fate of perceptions and sensations. In the prior subchapter, we saw some of the ways that

[12] Crane uses the word "unconscious" so I am reporting it thus. In general, I will use the word "nonconscious" as to avoid any Freudian associations.

sensations play a role in perception. We will now look at the question of whether perceptions themselves can be thought of as only conscious, and then the related question of the status of sensations. The former issue has been the topic of much scientific research as well as philosophical analysis; the latter one falls, much like a hot potato, only to the philosophers. I will cover that in Sect. 2.4. First, then, the issue of perception.

The traditional position, of course, was that perception is conscious. This underwrote the view that there is the act of perception and then there is the process of cognition, and that the two procedures are mutually exclusive of one another. On this view, to reiterate, perception provides the basic fodder for thought, but thought *shouldn't* assist in the construction of perception (e.g., to *think* you see something you don't in fact see is a hallucination, which is an instance of failed perception), and perception is simply *unable* to assist in thought—sensory experience/perception is restricted to the embodied realm. Objects are to be seen in perception, and then thought is to use that data in abstraction and concept formation. The putative claim is that to perceive something is to know that you've perceived it.

Recent philosophical literature though has pointed out several problems with this (Prinz, 2010).[13] I briefly review two of these in order to demonstrate further reasons (some preliminary ones were laid out previously in Chap. 1. Firstly, why it is illegitimate to argue for perception as identical to conscious experience. Bill Brewer has argued that the content view erroneously claims that representational content is constitutive of experience, and that it fails to account for the fact that the conscious naming and experiencing of objects presupposes prior visual awareness, which itself is only partially attentional (Brewer, 2017). This seems to me like a very legitimate point, as is his point that representationalism in general fails to account for the particularity of perceptual experience (Brewer, 2008, 11). I would also add that much of that particularity is not consciously perceived. Secondly, a different argument, equally plausible, comes from Jesse Prinz: Not all of the perceptual environment is processed as attentional, as illustrated by instances of inattentional blindness (Prinz, 2011). An instance of the latter is the now-famous gorilla-walks-onto-the-ball-court example, where 50% of the observers who were attentively attuned to the ball game did not see the person in the gorilla costume walk onto the court.

Prinz's interesting argument for the consequences to the theory of perception that inattentional blindness poses are not though perhaps relevant to the issues here at hand. His point was to focus on the general issue of attention. This has had a landslide affect in philosophy and is now considered a de rigueur point;[14] he and others did successfully convince us that much of perception goes unnoticed. But inattentional blindness is though distinct from what I would call nonconscious perception, for the following reasons. I will first give two examples from ordinary experience and then proceed to lay out the scientific studies in favor of nonconscious perception.

[13] See also Ian Phillips' "Unconscious perception reconsidered" (Phillips, 2018, 471–514) especially his very thorough footnote on p. 471 listing other relevant research.

[14] See my discussion in Sect. 3.4.4.

Our ordinary experience shows us that we are often able to remember atomistic parts of perceptual experience that we had first not consciously noticed. Let me give two different examples. Police often question and re-question witnesses, encouraging them to remember things they didn't initially consciously notice—the color of the perpetrator's clothes, distinguishing bodily marks, etc. In this recounting, the witness goes back over the scene almost as though they were reviewing a photograph or a recording—playing and replaying the memory, looking closely into crevices for bits that weren't at first consciously noticed. This is distinct from inattentional blindness because in that case there was, literally, a blindness: Subjects did not notice the gorilla at all, even during attempts at recall. A second example would be the following. We can often notice things that we hadn't consciously or originally noted in an experience upon a second re-viewing of the scene itself—we are reminded that we saw it the first time, but we had not "noticed it" consciously. For example, we might re-visit a roadside vista and upon looking out on the vista note that there was a little town there that we had not consciously noted the first time, but then upon seeing it again remember that it was—vaguely—there the first time. 'Oh, yeah,' we would say, 'I remember that.'

2.3.2 In Science

Science too has chimed in here and confirmed the vast array of perceptions that are known nonconsciously. Some of these involve the subtle clues that pass between us in our daily interactions with others and that have a fairly profound effect on our social lives e.g., emotional stimuli that are often processed without being consciously perceived. Researchers, cleverly dubbing this "attentional unawareness", have confirmed that subcortical structures are important in this process, and furthermore, that these kinds of nonconsciously perceived emotional stimuli "induce distinct neurophysiological changes and influence behaviour towards the consciously perceived world" (Tamietto & De Gelder, 2010, 697). Subcortical structures are the areas of the brain that are involved in emotion, hormones, and memory and that significantly affect our social functions; some of the main anatomical areas included in this area are the amygdala, the hippocampus, the thalamus, and the putamen. By activating these parts of the brain, nonconscious perception is allowing us to react to each other without even consciously knowing what it is that we've reacted to.

And, actually, this is a good thing, and it makes sense. It is very important to us that we *behaviorally respond* to the minute emotional clues given by others' facial features, tone of voice, bodily stance, and gestures—extending to and including such things as habitual walking patterns and "at rest" facial expressions. It is not so terribly important that we know *to what* we are responding; what matters is that we have acted appropriately and protected ourselves; it is about survival. But let us return to the things themselves to which we behaviorally respond. All these things communicate—the minute movements of facial muscles seen in the tenseness of a smile, the narrowing of eyes, etc. And they are important to know: Knowing what someone feels is knowing their intentions. And we do let each other know our intentions. Who we are is trumpeted in a million ways. Or as the writer Maya Angelo once said, "When people show you who they are, believe them the first time." She was pointing to the fact that we *do* show each other, and we are surprisingly able to decipher this information, even without being consciously aware that we are. The difficulty, of course, – and this is what Angelo was pointing to—is trying to pull that nonconscious awareness into the realm of consciousness. But that is not the issue with which we are here dealing; trying to establish rules for how we gird ourselves with the armor of the conscious knowledge of others' intentions is not the purpose of this book. Instead, our current problem is trying to locate just *how much* of perception operates on the level below consciousness. For knowing that allows us to better analyze how it is that we see and absorb art—a fundamental kind of communicative perceptual process—and how it is that we derive meaning from that process.

Thus, Tamietto and de Gelder's research on attentional unawareness argues for the prioritizing of emotional information (ibid., 698) and confirms Gantman and Van Bavel's research discussed in Sect. 2.2.3 in regard to moral terms (see Gantman & Van Bavel, 2014). Tamietto et al. note the now widely recognized view that "Only a fraction of sensory input gives rise to conscious perception", and they make a further distinction between the two causes for this fact. (1) Selective attention, the result of both bottom-up and top-down saliency maps, gives us attentional awareness; but there can be selective awareness without attentional awareness as well. We can select, but not know that we've selected. This is the phenomenon they call "attentional unawareness". (2) Failure to consciously perceive can also be "due to sensory reasons": If the sensory stimulation has a presentation time that is "too brief (that is, subliminal), the stimulus often does not generate a conscious sensation even when we are paying attention to it. In this case we use the term 'sensory unawareness' (Tamietto & De Gelder, 2010, 698)." The latter, of course, is gist.

Like other similar studies, this study shows us, through the phenomenon of nonconscious perception, that it is not the case that perception is simply a mirror of reality; rather, reality is edited. If we, then, agree that reality is edited, the question becomes, "By exactly what mechanisms?"

2.3.3 Gist and the Nonconscious

Gist goes far in explaining this process. Let me reiterate some of the main points in this subliminal system of ours. The usual definition of 'gist' is what is given in the quick glimpse—the first impression—that one immediately has of a complex scene. Unlike object recognition, which is primarily controlled by the ventral occipital lobe, whole-scene gist—that perceptual experience that happens in milliseconds and that gives us a surprising amount of information—is activated primarily by the parahippocampal place area (PPA) and the retrosplenial cortex (RSC). This is activated in various situations. When one enters a room, one gets a gist of the room before one identifies specific objects, when one is presented with a vista of the outdoor landscape one sees the layout of the scene and whether it is calm or exciting, or when one is briefly shown a movie trailer one can recognize the owners of the faces. Although I have noted that much research has been done on the significant amount of semantic and perceptual information that observers can gather in the 200 ms that constitute a "glance", my interest for present purposes is in the qualitative component of the gist—and how it functions cognitively.

People are often not aware of their initial gist experiences, and yet it conveys vital information to the rest of the perceptual and cognitive processes. While objects are things that can be moved, and that we identify as things to be possessed, owned, or passed over to others, the gist of a scene gives us not objects but the general layout of the situation. Gist cannot be owned and it cannot be passed to others. There is no proprietary value to gist. You cannot even use it for a financial structured product such as a derivative! It is vaporous, slightly ethereal, and completely and ineluctably private. Yet it is crucial for us, as it is our first entry into the experience.

But it is not just the first entry, for it gives us more importantly *our reaction to* that situation: our quick assessment of liking or not-liking is captured in that first few milliseconds. It is the pull and push of perception: The desire and aversion of our bodies to the environment in which it finds itself. It is an important component of our emotional reaction. And that reaction of desire or aversion then influences the visual and cognitive processes that follow: When we initially like the gist of the scene, *then we look harder*—see more details—as we discern the constituent objects. We are also more willing to accept what we see. We parse the world differently. Gist determines our basic perspective and attitude toward the scene. It determines, at least in part, our editing.

Thus, gist is largely a phenomenon that is seen under the rubric of nonconscious perception. Recent research into the very early stages of perception have shown abundant amounts of information that we receive at this stage of visual processing, and it is now commonly referred to as "nonconscious" or "unconscious" perception. For example, one study showed that it is the early processing of the visual cortex that is useful for both conscious and nonconscious vision, while it is the late processing that is necessary only for conscious vision (Hurme et al., 2017). This particular research, like many others, focuses on blindsight patients e.g., those who have suffered lesions in their primary V1 visual cortex (sometimes referred to as the striate cortex), whereby their visual cortex is unable to "see" the perceptual information—with the patients claiming they can*not* see it—and yet the patients will, with statistical reliability better than chance, respond in appropriate behavioral ways *as though* they did, in fact, see it. Hurme et al. demonstrated that a physiological model (as opposed to a purely anatomical one) showed the more complex relations between the various parts of the perceptual system, noting that the feedforward activity from the V1 area to higher areas "mediates unconscious processes whereas feedback loops of recurrent activity from higher areas back to V1 support conscious vision (ibid., 230)". More specifically, they showed that it was the very earliest milliseconds of visual processing (60 ms) that was necessary for both unconscious and conscious vision, and that it was later processing (after 90 ms) where the V1 contributes selectively to conscious vision (ibid.).

There are two things to note here: (1) information is being guided by processes of which we have no awareness, (2) feedback loops play a significant role in perceptual processes. We cannot look at perception in the old-fashioned way the traditional empiricists did, with the arrow coming straight from the object, into the retina, and then culminating in a corresponding idea of the external world. This is simply not the way it works. The bulk of the work is not accomplished by the eye, but is done in the visual cortex, with many complex feedback loops; that is where the heavy lifting of the labor is. Therefore, when the blindsight patient has suffered lesions in the primary visual cortex, there are not the feedback loops necessary for conscious perception. And yet they "see" because the nonconscious perception has been able to occur.[15]

It is worth directing our attention to one last study in this regard and spelling out in detail the experimenters' process. In 2016, Chen Song and Haishan Yao published an article proving that unconscious processing of subliminal information occurred not only in blindsight patients but in normal ones as well (Song & Yao, 2016). A testing stimulus (low-luminance grayscale) was presented to one eye but that testing stimulus was rendered invisible by a masking stimulus (high-contrast colorful mask) presented to the other eye. There were three separate experiments, all designed to test the influence of luminance and complexity on subliminal vision, with the first experiment using a gray-scale sinusoidal grating that was either tilted 45° to the right or 45° to the left, the second one a gray-scale of a female face or a house, and third one a gray-scale image of a sad cartoon face or a happy cartoon face. The mask was a colorful Mondrian flashed at 33.33 Hz. The testing stimulus and the masking stimulus were presented for 300 ms after which the testing stimulus was replaced by the masking stimulus with the masking stimulus presented to the two eyes for 600 ms. Afterward, subjects made a forced-choice discrimination of the stimulus identity, having to choose between the alternatives in each of the three experiments (e.g., is it 45° to the right or to the left?, is it a female face or a house?, is it a happy or sad cartoon face?). In each experiment participants completed 1250 trials. Their results showed that "even without awareness of the stimulus, participants could nevertheless reach above-chance accuracy in discriminating the testing stimulus, regardless of its luminance or complexity" (ibid., 38920). They concluded, "These results suggested that participants could form unconscious knowledge of subliminal visual information, although not as accurate as the conscious knowledge of supraliminal visual information" (ibid.).

It is now uncontentious to claim that there is nonconscious perception. The last decade or two has seen much published in the scientific community regarding this, and we have reviewed several of these findings. What then does this mean for our investigation into perception and its role in art? I think it means several things. Let

[15] See also Lamme (2001).

me quickly present some obvious consequences of these findings by examining the effects on several older philosophical arguments.

First, the old-fashioned mimetic notion of aesthetics, in truth discarded long ago, suffers another blow. Art is not about reproducing the consciously ascertained "reality". Since our perceptual processes involve so much more than what we are consciously aware of, and furthermore since perception itself is the result of massive amounts of editing, art cannot be understood as having the job of replicating a non-edited and fully consciously perceived world. Such a world does not exist.[16]

More interestingly, the fact of nonconscious perception discards also a semantic, referential analysis of art such as that provided by Nelson Goodman. Using visual art as the paradigm, the semantic analysis is one whereby art is putatively understood through giving an analysis of how the image refers in semantic terms (Goodman, 1976). But it is important to note that semantic reference is accomplished *after* the fact of experiencing the art; it is a description of the correspondence between the artwork and what it is referring to in the world—a sort of matching endeavor, whether that be primary reference or metaphorical reference. In either instance, it is a matching function that the perceiver (or writer) is doing *after* looking at the image and then determining what it is referring to in the world. For example, one would look at a grey painting of the ocean and determine that it was referring literally to the ocean and also referring, metaphorically, through the grey color, e.g., to sadness.

But that is not enough of an analysis of art. What we want is some kind of account of *what happens at the moment we are standing* in front of the artwork: How do the multiple eye saccades that I have when I look—my eye darting here and there, repeatedly experiencing the gist of the scene, followed by moments of intense attention, and then followed again by the extremely fast darting of the eye—how does *that experience* translate for me? Most importantly, how does that dive into the low-level features—*which do not have semantic correlates*—translate into phenomenal experience? And, furthermore and fundamentally, how does that translate into meaning? This is especially problematic for Goodman as meaning was to be eschewed in favor of reference.

Thus, a semantic account, such as that argued for by Goodman, with all its pretend precision, cannot give us that. Visual art cannot convert, without significant loss of information, into semantic descriptions. There is no one-to-one correspondence. There is not even a loose or metaphorical relationship, especially if that relationship is being built upon the back of the false presupposition that looking at artworks is analogous to looking at linguistic objects. We perceive much more than linguistic objects conceived of as the semantic equivalent of the visual presentation i.e., the things in the world. In addition, there is also the point regarding meaning:

---

[16] One might counter by saying, "But look at photography!" Nelson Goodman makes some arguments against this (Goodman, 1976), but I think there are some better arguments in regard to this. Photographs are a very distorted and inaccurate view of the world as they represent all of the image as though it were foveated. All is in focus, none is peripheral. But that is absolutely not the way we see.

To claim that meaning can be given a reductivist analysis via reference is insufficient for the social complexity of the communicative functions of art. We need to do better than that.

Thirdly, an analysis of art in terms the "expression" of the physical object is likewise not satisfactory. Richard Wollheim's analysis is an apt example of this. It was he who was adamantly opposed to looking at art as an instantiation of some quasi-lofty goal of enacting ideals (e.g., beauty would be among those), and for that we can thank him. But his attempt to tie the experience of art (and the understanding of it) to a psychological analysis of artworks, (paintings and sculpture in particular), and to examine our experience in terms of expression, is alas not quite what we are after. It is not the object that is being "pressed out" (Wollheim & Wollheim, 1980)—with psychological associations expressed by the physicality of the object itself. Nor is it that art is an instance of "seeing-as" whereby the image is revealing of the representational value via the artist's intentions.

But in all fairness, these last few paragraphs were an exercise in low-lying fruit. There have been significant criticisms of these older twentieth century versions of analytic aesthetics, both by others and by myself. In contradistinction to that, much of the last decade or so of recent philosophy that deals with art (either as aesthetics or as philosophy of art) has left the terrain of these formerly well-trodden and dead-end paths and has instead turned to an analysis within perception. This is of course the right approach. We cannot understand how the perceiver experiences art (or how the artist experiences—see Chap. 4) without understanding perception, and that includes the basic scientific studies that have given us the relevant empirical data.

Thus, and instead, we must look to art as an instance of gist perception and an instance of nonconscious perception, embedded in which is the communicative editing of the artist as that artist tries to convey a version of lived reality.

2.4 Sensing and Visual Art

2.4.1 Sensations and the Experience of Art

In the prior subchapter we examined the question of whether perceptions themselves can be thought of as only conscious, concluding that they cannot. We also noted that there was the related question of the status of sensations—how these then integrate themselves into the role of both conscious and nonconscious perception. We now come to the treatment of sensing in visual art, which necessarily brings with it the related role of sensation.

There has been often in the history of philosophy a separation between the things themselves and the sensations we associate with them. There is an apt 1935 Heidegger quote that is often pulled out in this context:

> We never…originally and really perceive a throng of sensations, e.g., tones and noises, in the appearance of things…; rather, we hear the storm whistling in the chimney, we hear the

> three-engine aeroplane, we hear the Mercedes in immediate distinction from the Volkswagen. Much closer to us than any sensations are the things themselves (Crane, 2006, 126).

His "throng of sensations" highlights what is often called the division between the representationalists and the qualia theory philosophers. I am not going to address this debate directly, as our concern is focused on the role of sensation in the experience in art. But, as we shall see, once we address the role of sensation in art, these other issues will be at least partially reframed and clarified. Thus, we turn to the issue with art.

To reiterate, our question is: *what happens at the moment we are standing* in front of the artwork? More specifically, how does sensation, which is involved in perception, filter into the process of experiencing the artwork? It is here where the rubber hits the road. As one of the most difficult questions in philosophy, it is also an enormously worth-while one. Human art-making and human willingness to experience other people's art is very distinctive to our species. I would argue it is *the* distinctive trait. It used to be said that humans were distinguished from the animal world through the exercise of speech, but as research in the expanded categories of "communication" and "language" has ballooned, the sacrosanct notion that it is only humans who "speak" to one another is no longer thought to be true. As the cognitive biologist Tecumseh Fitch has noted, "My research has shown that in fact that most aspects of human speech, including the anatomical, acoustic and neural bases of speech, are shared with other nonhuman species."[17] So, it is not just that animals merely vocalize while we humans speak, but that our speaking seems to not be founded upon structures that are singularly ours; we share more than we thought and likewise are less special than we thought. A related point in the history of thought was the claim that the distinctive human trait was rationality, which, being linked to verbal speech, was another contender for the special gift owned only by humans, thereby allowing humans claim on a higher rung of creation. But that claim too seems diminished as a consequence of the significant work now devoted to animal cognition (Lurz, 2009).

But art is something different. Art really is something that only humans do; we have never seen a dog pick up its paw, draw some lines in the dirt, and stand back admiring it with other dogs trotting over, joining in, and rolling their eyes or howling. But we do things like that—well, more or less. So why is art singularly human? And what do we get out of it?

In order to continue our investigation into this, we must be specific about what happens the moment we are in front of an artwork. More specifically, if we—for example—look at red in a painting, how is that different from looking at red in a department store while one is perusing the racks? For, surely, our activity of looking for a red shirt in the department store is not significantly different from what all animals do when they peruse a scene in hopes of finding food, water, shelter, etc. Looking at red in art though is different, and articulating that difference will allow

[17] https://homepage.univie.ac.at/tecumseh.fitch/wp-content/uploads/2011/05/FitchCV2011.pdf/

us to understand what this distinctively human activity called "art" is, and to see the role sensation plays in the perception of art.

2.4.2 Property Attribution Versus Gist

We can begin to develop a solution by introducing the notion of distributed attention. This notion distinguishes itself from focused attention, which is to focus on one part, one singular entity. On the other hand, to distribute is to do exactly what you'd think: It is to distribute one's attention over the physical space of the object or scene in question. This model has been well established in psychology for many years (Srinivasan et al., 2009).[18] We very often distribute our attention. We have noted the fact of eye saccades—the rapid movement of the eye four or five times every second—as well as the role of saliency maps, which govern what is selected from the plethora of data that is possible to choose from. Similarly, when our eyes scan a real-life scene or a painting on the wall, it is not foveating on a specific part, but rather the eyes are rapidly moving across the surface. I will discuss more of this in a bit.

But before I do we need to remember that before the scanning or distributed attention begins there was the initial gist global-scene experience of the whole of the painting: We take in the whole first. And then we can either focus our attention and foveate on a singular part, or we decide to distribute our attention and scan—rescanning again and again different individual parts of the whole. This is what we do: We glance very quickly at the painting, taking in the whole of it (e.g. whole-scene gist), and then move a bit closer and scan, scan, scan, eyes rapidly moving this way and that, looking at different moments and movements in the plasticity of the picture. Or, alternately, we glance very quickly at the painting, taking in the whole of it, and then we only give focused attention to, for example, the face in a figurative painting, ignoring all other parts.

But for the purposes of our argument since we are trying to articulate the notion of distributed attention, let us say we are looking at the painting in a distributed way by scanning—glancing at different parts letting our eyes roam over here and then over there. And, furthermore, let us say this picture has a significant amount of red in it, such that the perceiver's eye is rapidly moving between the distinct instances of red, darting over here and then over there, while all the time maintaining whole-scene perspective. What is the perceiver doing in this example—in this aesthetic instance—that is different from what they did the day before at the department store while looking for a red shirt?

The answer is in a further sub-distinction e.g., that between two different kinds of distributed attention: (1) property attribution, and (2) whole-scene gist perception seen in very quick eye saccadic moments. I will give a basic definition and

---

[18] See also (Nanay, 2016).

over-view of the latter, which was briefly described two paragraphs ago, and then proceed to the former.

Gist thus occurs in two different ways when we are experiencing an artwork: (1) in the first whole-scene glancing of the artwork itself, and (2) in the individual saccadic scans that also provide discrete instances of gist perception. The latter are mini-global scene gist experiences. For example, the eyes have darted over to a little blob of red and thus the eyes have truncated their parameters and narrowed the range of information to a minute singular low-level moment in the painting e.g., this blob of red. But that few milliseconds of red-perception share its basic phenomenology with the first global gist scene perception of the painting itself.

It shares that phenomenology because in both instances the limited milliseconds of perception provides both low-resolution information, which is dependent upon low-level features, and it provides our basic subjective reaction to the data. Thus, it is giving us both the objective data e.g., "it is a bit of red", as well as the values derived from socially constructed choices e.g., it is violent/passionate/emotionally-gripping, etc.

Gist thus gives one's own bodily reaction to that fact: Like or dislike, yes or no, desire or aversion. It is our bodies' reading of the environment and is an essential ingredient to the emotional component of experience. Thus, alas, Hume was wrong about this one: Value is not "added on" to empirical experience; it is an integral part of it. This is one of the important consequences to gist research.

Thus, these minute gist experiences of the interior marks of the painting give us enormous amounts of information. They accomplish this because they are numerous and cumulative, with some functioning consciously and some nonconsciously. Furthermore, and importantly, it is these that provide the underpinnings for the experience of what we call "sensation". I will explain the physiological mechanisms of this more in detail after the promised brief analysis of property attribution.

The usual definition of property attribution is that it is the ordinary kind of perceptual process, such as the one in the department store. One is scanning, looking for red, in such a way that the *red belongs to* the shirt; the red is a property that is attributed to that shirt. This kind of relationship is the ancient philosophical coupling of the object and its properties—for Aristotle, the red shirt has as its predicate "red", for Plato the red shirt instantiates the universal property of red. Thus, there are many ways to express this: The red is possessed by the shirt, the property red is instantiated in the shirt, the predicate red is attributed to the object shirt. Regardless of how it is phrased and whatever metaphysics are being asserted, property attribution is the skeletal relationship on which semantic truth is based.

What does it mean to attend to properties? Do we attend to properties—the multiple tokens of the type red—in a gallery setting in the same way that we attend to the properties of red tokens while looking for a red shirt in the department store? Intuitively, one wants to say "no". In the department store, we are, it is true, scanning and our attention is distributed; in this way, it is like the aesthetic experience. But in the department store our attention is directed toward the identification of that property *as belonging to that object*: the red belongs to the shirt. The red was targeted, named, and linked to that object. There is a metaphysical point here as well as the epistemological one. In the activity of searching for a red shirt, *the entity "shirt" has already been ontologically chunked, and its predicates assigned.* The property belongs to the object, and we are using the property in order to target the object. Again, we see that the relationship of property attribution is the basis of semantic content.

But we are not concerned with establishing the conditions of satisfaction for semantic content. Instead, we are concerned with articulating what cognitive functions accompany the moment we are standing in front of a painting and experiencing that painting. Gist perception while scanning an artwork is not at all like property attribution. To see red in an aesthetic context is to experience the distributed perceptual event as a gist experience. Much of this comes down to the mechanics of our eyes. Because it is only the small area (2–3%) of the retina that is the foveal area and that thus focuses, we must shift our eyes in order to point our eyes to our target. That pointing is usually not static; rather, in most instances our eyes are performing rapid eye saccades, each one lasting anywhere from 20 to 200 ms. This is the scanning that we constantly do. And this is what we do when we come upon a painting and allow our eyes to dart around its surface, seeing the red here and there and again over in that part.

So, it does not seem plausible that we perform, in aesthetics, the same task of finding and checking off the presence of property attributes in the object. I did not, in the gallery, go around the room and say, "There's a red, there's another red, and here is yet another one! These two match!" It is obvious that *that* is not what we are doing. *We are not looking for the reds in the painting in order to identify them as a property belonging to (x).* We are not *looking for* and semantically attributing properties. Distributed attention of the gist variety is not an instance of property attribution.

2.4.3 Attentive Versus Pre-attentive

What is it, then, that we are doing when we experience aesthetics? Words can sometimes lead us in the right direction, and in this example that is true: We *experience* red in the aesthetic setting. Let us examine what that means, and briefly, let me summarize what I will argue. I will argue that it is the examination of distributed, sequentially experienced pre-attention (i.e., gist perception seen in very quick eye saccadic moments) that shows us the role of low-level features in the perceptual experience of an artwork. Furthermore, these low-level features are singular entities that are often non-linguistic and sometimes nonconscious and are themselves what we usually call "sensations". They are the lowest ontological level, what I am calling P-unitatas.

To begin, let us reiterate that gist is distinguished from property attribution. A similar kind of division can be seen when we remember that vision scientists have divided perception into a "preattentive" phase and an "attentive" phase (Koch & Ullman, 1987, 115). This parallels the divide between gist and property attribution.

Property attribution is the conscious, attentive recognition of a property belonging to an object, while the predication associated with gist can be accommodated in the preattentive phase. Predicating would encompass those instances that do not rise to the level of *F of x*. For example, I can experience the pain in my leg but not be aware of it; but to *be aware of it* is to consciously attribute the property of pain to my leg e.g., as when I would say, "The pain is in the leg". Or, to give another example, I can experience the predicate red in the couch and be able to, if queried later, call it as such, but that is a very different perceptual experience than the instance of visually targeting the red couch itself *after* being told to sort the red couches from the blue. Semi-consciously or nonconsciously noticing it in the first place is different from consciously identifying "that couch is red" upon command. The latter process is highly attentive and, in addition, relies fundamentally on high-level semantic features.

To simply predicate though is different and this is analogous to that that is seen in gist perception: The property is noticed but the perceptual function is not to assign it to an object; this is why low-level features are so central with gist. *They function independently of object possession.*

Exactly how does this process occur? As was noted in the prior chapter in the larger discussion of saliency maps, our eyes are a mechanical, computational mechanism that absorb the external world through a series of sampled episodes driven by interposed and extremely fast saccadic scans. The samplings occur at 3–4/s; this makes for the obvious, yet often ignored, fact that a large proportion of our visual surroundings are not processed by us (e.g. Ballard et al. 1992; Land & Hayhoe 2001; Land, Mennie & Rusted, 1999; as cited Tatler et al., 2005, 643). The obvious question arises: What distinguishes the saccade target selections from those that are not targeted?

The answer is found in the distinction between "attentive and "preattentive". This was discussed in detail in the first chapter, and it is worth reiterating and summarizing some of the main points. (1) This distinction arose at the beginning in the 1980s, wherein many studies began to appear that characterized visual perception as two-stage process: The first stage of "preattentive" mode, in which simple low-level features were rapidly processed in parallel over the entire field, and a second "attentive" mode, where the focus was to particular locations in the visual field (Koch & Ullman, 1987, 115).[19] (2) A significant number of researchers who have emphasized the bottom-up role of "salience maps", have singled out for emphasis the activity of low-level features in those salience determinations (Itti & Koch, 2000; Itti et al., 1998; Olshausen et al., 1993). "Saliency is independent of the nature of the particular task, operates very rapidly and is primarily driven in a bottom-up manner…" (Itti & Koch, 2001, 194). (3) When determining what the oculomotor system prioritizes as salient, researchers have found that it highly influenced by context (ibid., 197), which is largely a mechanical response e.g.,

[19] Some recent studies though have questioned this determining influence of attention. See Rosenholtz et al. (2012), and my discussion of it in Sect. 3.4.4.

responsive to contrasts. This gives us a model of vision that is initiated by a bottom-up, semi-mechanical oculomotor system governed by relatively uniform standard of saliency that is then, in both a feedforward and feedback loop, interactive with top-down object-recognition processes.

It is plausible to also assume that context decisions are modulated by prior selections; the inductive practices controlled by constructed social preferences must also come into play. We would, for example, notice a pink patch in a painting more rapidly than a red one if we were, for example, active in the breast cancer movement which has pink in its marketing identifier. Remember the results from the researchers Gantman and Van Bavel that confirmed we see what we value.

In other words, the initial preattentive, bottom-up oculomotor process phase was generated more uniformly, and it was combined with—and contrasted to—the later top-down, feedback, cognitive, semantic, and object-recognition phase, the latter of which varied more from individual to individual. This shows that the initial low-level phase is generated more uniformly and thus governed by the mechanics of biology, whereas the later phases can be influenced more by the inductive practices of both the individual and the socially constructed patterns of linguistic habit and social norms. We see this pattern also in the process of looking at art pieces.

This is especially true with abstract art. Gist experienced at <100 ms. has been shown to not only provide the basics of color, shape, texture, etc., but has also been shown in studies to be fundamental in the understanding of abstract artworks. Subjects were shown images of abstract artworks in a masked exposure of 50 ms, and then asked to judge whether the artwork's composition was "stable, orderly, and harmonious" or whether it was "unstable, disorderly, and unharmonious" (Schwabe et al., 2018). These judgements regarding the stability and orderliness of the composition, judged at a mere 50 ms, did not change when the subjects were given the opportunity to view it at longer intervals of 3000 ms. In other words, the 50 ms judgments were consistent with those made at 3000 ms. This proves that subjects were attentive to the low-level features, and that those low-level features, extracted at 50 ms of their experience, provided the basis for objective assessments that determined the subjective sense of order/safety or disorder/unsafety. The affordances information was thus there encapsulated in the original low-level perception at 50 ms. (For further discussion see Sect. 3.4.4.) And that provided for well-founded judgements based on those experiences. Again, we see how the so-called Humean "value" is not added on to the empirical data. Rather, it is there from the beginning. And it is found in the role of low-level features.

This is central to our concerns, for it points to the conclusion that what is significantly constitutive of an aesthetic experience is its focus on low-level features, and contra-positively its ignoring of higher-level features. To put it in more philosophical jargon, we may say the low-level features that the aesthetic experience is focused on is pre-conceptual. In this, it doesn't necessarily involve the process of attributing properties to objects, it just requires the awareness of such features; it is also thus significantly pre-attentive.

References

Avecillas-Chasin, J. M., Levinson, S., Kuhn, T., Omidbeigi, M., Langevin, J., Pouratian, N., & Bari, A. (2023). Connectivity-based parcellation of the amygdala and identification of its main white matter connections. *Scientific Reports, 13*(1), 1305.

Ballard, D. H., Hayhoe, M. M., Li, F., & Whitehead, S. D. (1992). Hand-eye coordination during sequential tasks. *Philosophical Transactions of the Royal Society of London. Series B: Biological Sciences, 337*(1281), 331–339.

Brewer, B. (2007). Perception and its objects. *Philosophical Studies, 132*, 87–97.

Brewer, B. (2008). Perception and content. In *John McDowell: Experience, norm, and nature* (pp. 15–31). Wiley.

Brewer, B. (2017). Consciousness and content in perception. *Philosophical Perspectives, 31*, 41–54.

Chariker, L., Shapley, R., & Young, L. (2016). Orientation selectivity from very sparse LGN inputs in a comprehensive model of macaque V1 cortex. *Journal of Neuroscience, 36*(49), 12368–12384.

Clark, A. (2015). *Surfing uncertainty: Prediction, action, and the embodied mind*. Oxford University Press.

Crane, T. (2006). Is there a perceptual relation. In *Perceptual experience* (pp. 126–146). Oxford University Press.

Crane, T. (2013). Unconscious belief and conscious thought. In *Phenomenal intentionality* (pp. 156–173). Oxford University Press.

Evans, G. (1991). *The varieties of reference*. Oxford University Press.

Gantman, A. P., & Van Bavel, J. J. (2014). The moral pop-out effect: Enhanced perceptual awareness of morally relevant stimuli. *Cognition, 132*(1), 22–29.

Goodman, N. (1976). *Languages of art: An approach to a theory of symbols*. Hackett publishing.

Hurme, M., Koivisto, M., Revonsuo, A., & Railo, H. (2017). Early processing in primary visual cortex is necessary for conscious and unconscious vision while late processing is necessary only for conscious vision in neurologically healthy humans. *NeuroImage, 150*, 230–238.

Itti, L., & Koch, C. (2000). A saliency-based search mechanism for overt and covert shifts of visual attention. *Vision research, 40*(10–12), 1489–1506.

Itti, L., Koch, C., & Niebur, E. (1998). SHORT PAPERS-Computer Vision and Image Analysis-A Model of Saliency-Based Visual Attention for Rapid Scene Analysis. *IEEE Transactions on Pattern Analysis and Machine Intelligence, 20*(11), 12–1120.

Itti, L., & Koch, C. (2001). Computational modelling of visual attention. *Nature Reviews Neuroscience, 2*(3), 194–203.

Koch, C., & Ullman, S. (1987). Shifts in selective visual attention: Towards the underlying neural circuitry. In *Matters of intelligence: Conceptual structures in cognitive neuroscience* (pp. 115–141). Springer.

Lamme, V. A. (2001). Blindsight: the role of feedforward and feedback corticocortical connections. *Acta Psychologica, 107*(1-3), 209–228.

Land, M., Mennie, N., & Rusted, J. (1999). The roles of vision and eye movements in the control of activities of daily living. *Perception, 28*(11), 1311–1328.

Land, M. F., & Hayhoe, M. (2001). In what ways do eye movements contribute to everyday activities?. *Vision research, 41*(25–26), 3559–3565.

Lim, S., Padmala, S., & Pessoa, L. (2009). Segregating the significant from the mundane on a moment-to-moment basis via direct and indirect amygdala contributions. *Proceedings of the National Academy of Sciences, 106*(39), 16841–16846.

Lurz, R. W. (2009). *The philosophy of animal minds*. Cambridge University Press.

Nanay, B. (2016). *Aesthetics as philosophy of perception*. Oxford University Press.

Nietzsche, F. (2005). On truth and lies in a nonmoral sense. In *Truth: Engagements across philosophical traditions* (pp. 14–25). Wiley.

Noon, M. (2018). Pointless diversity training: Unconscious bias, new racism and agency. *Work, Employment and Society, 32*(1), 198–209.

Oliva, A., & Torralba, A. (2006). Building the gist of a scene: The role of global image features in recognition. *Progress in Brain Research, 155*, 23–36.
Olshausen, B. A., Anderson, C. H., & Van Essen, D. C. (1993). A neurobiological model of visual attention and invariant pattern recognition based on dynamic routing of information. *Journal of Neuroscience, 13*(11), 4700–4719.
Parkhurst, D., Law, K., & Niebur, E. (2002). Modeling the role of salience in the allocation of overt visual attention. *Vision research, 42*(1), 107–123.
Peacocke, C. (1992). *A study of concepts*. The MIT Press.
Phillips, I. (2018). Unconscious perception reconsidered. *Analytic Philosophy, 59*(4), 471–514.
Prinz, J. (2010). When is perception conscious. In *Perceiving the World: New Essays on Perception*, 310–332.
Prinz, J. (2011). Is attention necessary and sufficient for consciousness. In *Attention: Philosophical and psychological essays* (pp. 174–203). Oxford University Press.
Rosenholtz, R., Huang, J., Raj, A., Balas, B. J., & Ilie, L. (2012). A summary statistic representation in peripheral vision explains visual search. *Journal of Vision, 12*(4), 14.
Ross, W. D., & Smith, J. A. (1910). *The works of Aristotle*. Clarendon Press.
Ryle, G. (2009). *The concept of mind*. Routledge.
Schellenberg, S. (2016). Perceptual particularity. *Philosophy and Phenomenological Research, 93*(1), 25–54.
Schwabe, K., Menzel, C., Mullin, C., Wagemans, J., & Redies, C. (2018). Gist perception of image composition in abstract artworks. *I-Perception, 9*(3). https://doi.org/10.1177/2041669518780797
Shottenkirk, D. (2009). *Nominalism and its aftermath: The philosophy of Nelson Goodman*. Springer.
Song, C., & Yao, H. (2016). Unconscious processing of invisible visual stimuli. *Scientific Reports, 6*(1), 38917.
Srinivasan, N., Srivastava, P., Lohani, M., & Baijal, S. (2009). Focused and distributed attention. *Progress in Brain Research, 176*, 87–100.
Stalnaker, R. (1998). What might nonconceptual content be? *Philosophical Issues, 9*, 339–352.
Tamietto, M., & De Gelder, B. (2010). Neural bases of the non-conscious perception of emotional signals. *Nature Reviews Neuroscience, 11*(10), 697–709.
Uzgalis, W. (2007). *Locke's 'essay concerning human understanding': A reader's guide*. Bloomsbury Publishing Plc.
Wollheim, R., & Wollheim, R. A. (1980). *Art and its objects*. Cambridge University Press.

Chapter 3
Seeing: *A New View*

Abstract Given that whole-scene and gist perception (including peripheral, low-resolution, low-level, and nonconscious perception) are central to our perceptual faculties and depend primarily on both neurological constraints and probabilistic processes instead of conceptualization, this chapter argues against a role for concepts in (1) representation in art (2) conceptual art (3) perception theory. Coupled with these rejections is a rejection of the feedforward model (agreed upon by both the conceptualists and anti-conceptualists), replaced by feedback models that depend on loops and thus create dynamic systems. I argue that perception cannot be thought of as non-cognitive, and in fact the non-sensical notion of non-cognitive is itself to be discarded as well as the putative divide between perception and cognition. Instead, gist and sensation, which significantly accounts for the "what it's like" of experience, are counted as part of perceptual cognition—an embodied process. With a seamless relation to cognition, gist is a dynamic modeling of both the external world and internal affordances. The brain processes the data from the repeated gist perceptual experiences, which in turn can become constituents of further processes. The divide is between sensing and naming, not perception and cognition. Art, as paradigm, mimics the reiterative process of gist perception.

Keywords Feedforward models · Feedback models · Conceptual art · Cognition · Nonconceptual · Lawrence Weiner · David Marr · Non-cognitive · Naming · P-unitates

3.1 Introduction

We are now at a point where we can knit together the threads of the conversation and give a whole cloth description of the perceptual experience in general, ending with the more specific experience of standing in front of an artwork. This involves points covered in the previous two chapters: That we go from sensing to naming, the role of affordances and emotional reactions in the gist experience, the causal and

D. Shottenkirk, *Art as Cognition: How Gist Reframes the Aesthetic Experience as Conversation*, Synthese Library 521,
https://doi.org/10.1007/978-3-032-00143-6_3

functional role of gist, the fact that all information comes already embedded in a schema and hence with meaning, the fact that perception is not passive and that editing is central as seen in the related roles of saliency maps and the iterative nature of the gist experience in eye saccades, the fact of the fine-granularity of particularities (e.g., p-unitates) that give us the "what it's like" feature of lived experience, the fact that we edit by attending to things that we've had past experience with or things that we value, the fact that perception also involves non-conscious experience, the distinction between property attribution and gist experience, and last but not least, the incredibly large role played by low-level features in gist perception.

All of this combines to force us into concluding that perception is much more than of linguistically-defined objects. Whole-scene and gist perception, including peripheral, low-resolution, low-level, and nonconscious perception, are central to our perceptual faculties.

3.2 Low-Level Features, Not Concepts

And this in turn forces us to reconsider the role of concepts. The usual view is that it is semantic concepts that delineate the divide between low-level and high-level features. The basic points are as follows. We apply concepts to things in the world e.g., we identify the object as a "tree": We've taken the mental notion of the general term "tree" and applied that to the specific instance instantiated in front of us. Concepts govern the semantic and provide for conditions for satisfaction for the generality constraint. This much seems generally accepted; it is the usual view.

It has also been generally assumed that veracity lies within the domain of concepts: They are the harbingers of truth; to know the general notion of "tree" is to know a truth more elevated and useful than knowing that the specific instance before you is called "a tree".

But let me not give the impression that the "usual view" is held absolutely univocally. While much is held to be uncontentious, there is much else about the specific roles and definitions of concepts that have divided philosophers. While that more detailed conversation is not within the purview of this book, a pointed conversation about the more specific application of concepts in the domain of art is pertinent. The argument for this conversation is as follows.

Firstly, in order to understand the notion of cognition within philosophy, one must understand the role that concepts play in that. It is in cognition that concepts are said to be applied as categorizing devices for entities (objects) provided through perception. Secondly, in order to understand the notion of high-level features in vision science, one has to understand the role that concepts play in that. It is in high-level features where concepts guide and categorize the low-level. Thus, concepts are central to our investigation.

But before that gets addressed it is important to further clarify any possible misunderstandings in my use of these terms. The two terms i.e., high-level and concepts, do not refer to the same thing. This is the case not only because the former is used in the field of vision research while the latter is in philosophy; it is also the case that the former refers to the identification of general semantic types *within the perceptual process*—the invariant version of the object—while the latter philosophical term is referring to a mental representation or an ideal abstract object and is located not in the perceptual process per se but in the cognitive process.

But, yet, a relationship is there. For it is with the aid of mental concepts that one can apply the category to the invariant object. And in both cases, it is a type/token relationship. And in both cases, they are categorizing methods. Thus, while it would be wrong to treat the two terms as synonymous, it would also be wrong to not recognize the related nature of the terms, particularly as we examine the usual, standardly accepted relationship between perception and cognition as well as their individual definitions.

I will take the reader through three different problems, each of which have devoted to it a section. (1) The problem of concepts as they are used in the philosophical tradition of emphasizing representation in aesthetics, (2) The misunderstanding in philosophy regarding the use of the term "conceptual" as used within conceptual art, (3) The problematic role concepts have played in perception theory.

3.2.1 Putative Concepts in Representation

A few paragraphs back I had repeatedly referred to the "usual, standardly accepted relationship between perception and cognition". This was perhaps disingenuous as I'm going to contend with it. I will go through some of my worries and reservations about this issue in this sub-chapter, which are meant to criticize the outlines of this problem and set the stage for the further investigations in Sect. 3.4.4.

Even the usual, uncontentious view has had a profound (and negative) effect on aesthetics, for concept-driven objects have been too-often thought to be the central feature in the experience of art.[1] Aesthetic theory often approaches art in terms of representation issues, wherein the art is typically described in terms of what the images are referring to. Thus, objects (both animate and inanimate) loom large.

This should not come as a surprise. Linguistic objects, understood as instances of general terms and the concepts that delineate them, are usually taken to be the way we see the world. This is both the everyday norm as well as the philosophical one. The latter view holds that concepts govern the semantic and provide the conditions for satisfaction of the generality constraint; it is concepts that allow us to categorize specific objects and thus give us an object-centered world. Accordingly, we have Heidegger's example, where we never originally experience the sensations themselves but the things that accompany the sensations i.e., "the storm in the chimney".

[1] This problem is more evident within philosophy than within either art history or art criticism.

Correspondingly, art is often understood in terms of the subject-matter portrayed; for example, it is a painting of a chimney.[2] The concept "chimney" is used to identify its referent in the world i.e., the (represented) chimney before us. Thus, to understand art was to understand it in terms of its representation: In terms of things, in terms of objects, in terms of the subject-matter—and thus the concepts that govern them. This is true even when the aesthetic theory relied on notions of expression or psychology, for it was still explained in terms of the representation and imagery, denoted through concepts.

But this limits us radically and is not the preferable way of parsing art, as it makes it difficult to not only parse the meaning of abstract or even conceptual works[3]—a very serious problem if one wants to think of any art after 1900—but to understand the epistemological function of art itself. For if we think of art as mere representation and symbolism, and *are relying on the role of reference*, then it is impossible to explain the difference between two paintings that are portraying the same thing, or the difference between identical iconography in a graphic-designed advertisement on the subway and the same iconography in a painting in the museum, or the difference between one painting of a chimney that is terrible and the another one that is great. And last but not least—it is impossible to explain the central function of gist perception.

We must, instead, take into account (1) the communicative value of the low-level features and (2) the role of context. (Context is also discussed in Chaps. 4 and 5.)

This then opens the door to *the way* something is depicted, and not just to *the what* that is being depicted. Reference does not rule. To know what something is referring to is not the central part of visual art and not much knowledge is gained by focusing on that. Instead, we must have heuristic systems that allow for the consideration of the low-level things themselves e.g., the textures, the illumination, the composition that results from low-level resolution, the orderliness, the levels of contrast, etc.

For it is true in general that all visual art pieces tell us not just (or primarily) the linguistic referent of their subject matter, but *the attitude toward that thing*[4]—e.g., the relation with which we ought to stand to the thing. Or in the words of the contemporary poet Tishani Doshi, "Language is never just about what is being said; it is also about how it is held" (Doshi, 2023). This is perhaps even more true once we leave the terrain of poetry and enter visual art where words with their referring capacities are absent. So, while we might concede that concepts divide low-level

[2] Correspondingly, there would be no paintings of a storm qua storm, and the counter-example of storm-clouds would not solve the problem for that too would not be the storm qua storm.

[3] This might not seem obvious at first. But much conceptual art relies on the formal properties (i.e. significantly the low-level features) underlying the idea even more than on the idea itself. Thus, the reliance on minimalist forms, grey scales instead of colorful palettes, or a monochromatic presentation. Think of the early artists such as Robert Barry, Christian Boltanski, Mel Bochner, Sol LeWitt, or Lawrence Weiner. I say more about this in a few paragraphs.

[4] I owe this to a comment by Carolina Flores when she was a Resident Philosopher in my talkPOPc project.

features from high-level features, we don't have to concede the traditional importance afforded to the role of concepts themselves within aesthetic theories. They don't signal the referent as important because the referent isn't important: Representation is a misguided method by which to interpret art.

3.2.2 *Putative Concepts in Conceptual Art*

A few paragraphs back I cautioned against using concepts in order to understand conceptual art. I imagine more than a few readers guffawed. Ding, ding, let us take our corners.

I think it is common knowledge that one of the key things in art criticism is being sensitive to *the way* something is done. This is entirely and extremely distinct from *what* is being done. The angels are in the details, and it is in noticing the subtleties and contradictions where one finds meaning, and so a good critic or art historian has learned these skills, as well as learning that relatively little of visual art is constituted by either a semantic description of the subject matter (the linguistic objects depicted e.g., a Christ figure, or a sexually alluring woman, or a monster, to name the top three contenders) or by the concepts themselves that are illustrated (e.g., the concept of salvation, or sexuality, or terror). For those kinds of descriptions are too broadly construed to address the specificity of the piece before one. Let me provide an example: Learn to Read Art (Fig. 3.1).

Here is a piece by the early conceptual artist Lawrence Wiener: "Learn to Read Art". First, let's start with the most obvious gist perception component: the colors,

Fig. 3.1 Lawrence Wiener. *Learn to Read Art,* 2019. 18 × 18 in. Screen Print on cloth. (Permission of Printer Matter, Inc.)

As noted previously, this can be seen at 42 ms. So, this is one of the things we would note first, and the meaning of those colors would have immediately started to guide our interpretation.

And these are very specific colors. The red is not a cadmium bright right nor is the blue an intense cobalt. Instead, they are both shifted slightly to a lighter hue with the addition of white, making them more demure. This both reduces the intensity and passion of the image and also increases its banality—our eye can glide over it without being stopped in its tracks; we can look and ignore at the same time. Next in obviousness (although at much larger milliseconds), the spelled-out sentence is an imperative. Like instructions to a child, it is a blanket order, an almost rude admonishment. But this rudeness is contradicted by the cheery choices of color as well as by the design: That element is very ordered with the box of X's, but that order is mixed with the not-ordered and slightly off-kilter presentation of the typeface. Because the latter is at a slight angle, this gives a lilting, light-hearted, and semi-comical reading to the piece. Angles, remember, spell disorder. But because the angle is slight, the disorder reads as energetic and exciting, not dangerous. (It is a bad thing for us as a species that excitement and danger are differences in degree not in kind…).

Note that this design element would have been part of the early gist perception close to the time of perceiving the colors. Furthermore, in order to understand the role these aforementioned low-level features are playing, you need to imagine it differently: The piece would look radically different if the font were level or the colors bright and aggressive. All of these tiny, minute choices are crucial. Artists consider these things,[5] and the observant, sensitive critic will notice these things, too. Finally, in terms of formal and low-level features, the font is sans serif, so it is without embellishment: It is telling you how it is, just the facts. It is easy to see this font immediately with the first glance and it requires no fine-grained or high image resolution. No delicacy of lines. No romance here. Now we add in the content: You are being told to learn to read art.

Hilarity strikes! Not only do you have to admit that such a thing is necessary—one must learn to "read" or understand art as much as one must learn to read a language or learn a science—but you are being told it with a chirpy insouciance that makes you wonder if the messenger isn't some misplaced, time-traveled schoolmarm from the 1950s! There is a complete contradiction of the seriousness of the meaning with the silliness of the presentation. "Here we go class, let's learn to read art!"

But note here: the end result *is* the art. And that is this: To ponder the seriousness of the meaning much more than one would have without the silliness of the formal

[5] Sometimes not entirely consciously. I have noticed in teaching painting that students will sometimes say "this part of the painting is not working" and they are not sure why; they know enough to have their "intuitions" alerted, but not enough yet to know the precise source of the problem. This is probably an instance of perceptual expertise that has multivariate sources such as whole-scene gist perception and the consequent ability to identify incongruity, implicit and explicit learning, and also the cognitive function of peripheral vision.

and low-level components of the presentation. (Profundity strikes by sneaking up…) Worth noting is the fact that there is another layer of meaning here as this piece is one that is printed on a napkin for an event at the artists books venue Printed Matter. So, there is "reading" in a few different ways. But in both of these shifting contexts, there is still the importance of the low-level features: The silliness and frivolity of them both undermine and underline the other meaning.

It is also worth noting that in this case, as is so often true, those low-level features overlap with what we often call the "formal" properties in art criticism e.g., the colors, the contrast, the spatial lay-out, the illumination and hues, etc. And these are the things you learn to look for when you are both studying how to make art as well as when you are studying how to look at art. (In other words: Learn to read art, dummy.)

Thus, conceptual art does not rely on concepts. It does not rely on identifying the instantiation of hierarchical abstract, general terms/ideas. It relies on low-level features just as much as does more traditional, non-conceptual works.

The difference lies in the fact that the intention of conceptual art is to take the viewer to the consideration of a particular thought, i.e., *a belief*, instead of to the more inchoate emotional or psychological meanings embedded in most instances of visual art.[6] But a focused, abstract thought about something one takes to be true—i.e., a belief—is not a concept. The sentence "learn to read art" in fact has several different concepts in it i.e., "art", "read", and "learn". And beliefs aren't the same thing as concepts. One could hold a belief about a concept such as "learn" but that belief would be in the form of "I believe the concept 'learn' applies to the specific instances of 'to learn to play the piano' or 'to learn to hate philosophy.'" In this case, one can have the propositional attitude toward that sentence "to learn to hate philosophy" in terms of a "disagreement with", a "belief in", or an "incredulity of", etc. Regardless, one's *attitude toward* the sentence is not the same as a concept. (In regard to that point, I do not maintain that concepts are mental representations and do not align myself with the language of thought view in philosophy that grounds thought as representational and structured.[7]).

Thus, the art is not about concepts anymore that my last night's dinner of salmon is about mollusks; concepts might be related in tangential ways to the notion of beliefs or propositions, just like mollusks share some traits with salmon, but to equate them is a mistake. Again, the point must be made. Conceptual art is doing what all art does: Conveying meaning.[8] In summary, what artists have meant by "conceptual" is not at all what philosophers mean. This is the loosey-goosey of jargon-stealing that artists blithely indulge in; the word is close to what they are doing, it sounds cool, and that's enough. Even more importantly, (and legitimately),

---

[6] For example, the *Mona Lisa* (to take THE example) elicits a feeling of attraction in response to the smile. That is not either a belief or a concept, although it has tertiary relations to both.

[7] For a representative of the language as thought view see Pinker (2007).

[8] In this way, it puts the process closer to what now is thought of as a Fregean notion of content, whereby the sense (meaning) is much more discriminating than the reference. It is worth remembering though that Frege identified the referent of the predicate as the concept.

the conceptual artist is taking the viewer to the meaning of the piece (which is this specific attitudinal relation to a thought) via the same route that all artists do: Namely, whereby the meaning is ensured by the use of specific low-level features. Thus, conceptual art does not rely on the philosophical notion of concepts i.e., those things that allow us to categorize. What is going on with conceptual art is quite different.

3.2.3 *Putative Concepts in Perception Theory*

A third problem is the role of concepts within the philosophy of perception as it impacts the debate regarding the definition of, and the relationship between, perception and cognition. There are, of course, many different approaches in the philosophy of perception,[9] and so I am only going to address the two most widely practiced and relevant approaches that directly reveal the relationship between perception and cognition: Representationalism and the conceptualists/anti-conceptualists debate. I will address the latter first as some has already been said about the debate between these two views.

3.2.3.1 Concepts' Role in the Conceptualism/Nonconceptualism Debate

Again, to reiterate the usual view: In the putative border between perception and cognition, we first perceive and then cognition takes over. Thus, it is feedforward for both viewpoints (i.e., the conceptualists/anti-conceptualists). (See Sect. 3.3 for a further discussion of the problem of the feedforward models in vision.) There is also another point of agreement: Whether one is a non-conceptualist who holds that percepts are without concepts (i.e., percepts are non-conceptual) and that concepts appear only in the cognitive mode, or a conceptualist who holds that percepts output at least some degree (there are differences here regarding how much) of conceptualized information, percepts are (by definition) non-cognitive.[10] Thus, in this they also agree: Percepts are not cognitions and thus they are non-cognitive.

But what would it mean to absorb some information *without* cognitively processing it? This is unclear. Perhaps instead of trying to define "non-cognitive", let us try to analyze brain processes in such a way that we discover what falls under the

[9] I leave out of consideration such approaches as doxastic theory, adverbial theory, disjunctive theory, or the older sense-datum theory. The role of concepts within each of these arises as well, but in those cases it is not as pertinent to the issue under consideration, which is the division between perception and cognition.

[10] The usual language used by conceptualists reveals this; it is said that percepts "output" some degree of conceptual information, which is a way of disclaiming that they themselves are fully constituted by conceptual information. I would argue that this language is bit vague, but for a representative account of this view see (Mandelbaum, 2018).

category of "cognitive". And then we can discover what is excluded from *that* category and thus know what "non-cognitive" means. At that point, we will know what it means to say percepts are non-cognitive.

The following is a list of those things that fall under the category of cognitive.

Implicit Memory Falls Under Cognitive It is uncontentious to claim that we can process recognizable information *and* be aware of it even though we have not (1) reflected upon it, or (2) conceptually categorized it. For example, children and animals presumably experience the world, process the information as recognizable and, in some fashion, are aware of it without consciously reflecting on their awareness *of it*. They don't think, "hmm, I just experienced a great meal; beef is my favorite." There is (presumably) no ruminating. It is also true that while they are aware of the thing before them, they do not process it as an instance of a larger category. They are not performing conceptual categorizing: "This thing in front of me is an instance of beef." There is also no conceptualizing.

And yet, still, we would want to say that they were absorbing information from the external environment, *and performing some kinds of cognitive processes* on that information albeit ones without conceptualizing or reflecting upon the (now past) experience.[11] This is obvious as there are many parallel processes occurring in the brain that allows both babies and animals to absorb the low-level features, recognize the scene or room as one they've seen before, as well as to recognize certain discrete objects. For example, the dog sees the food and recognizes it as such and wants it or recognizes the doorway of the veterinarian's office and doesn't want it, or the baby sees the bottle and wants it while seeing the mashed-up peas and doesn't. For even without conceptualization or reflection, the brain must—in order for the organism to survive—process a great deal of low-level as well as some primitive high-level categorizing (e.g., the dog with the food, and the baby with the bottle) that can be done without semantic awareness.

Thus, it is not necessary to say that the baby and dog are involved in abstract conceptual thought. For it is possible to *inductively recognize something* without the conceptual application of abstract concepts. This is the core of the distinction between implicit and explicit memory, as well as the distinction between the procedural memory system (also called "probabilistic classification", which includes habit-based learning such as riding a bike and playing an instrument) and the non-procedural, declarative memory system (which includes both semantic and episodic).[12] This is, notably, also the way Large Language Models (LLM) such as ChatGPT or Bard work. These sorts of probabilistic, inductive models that predict the next word are certainly not (not yet, anyway) conceptualizing.

---

[11] See David Rosenthal's notion of Higher Order Thought (Rosenthal, 1993).

[12] See Shottenkirk and Chatterjee (2010).

Thus, we have now two separate facts: (1) there are instances of "cognizing" that are not "conceptualizing", and (2) recognizing is not necessarily a function of conceptualization. This kind of recognizing we can refer to as "implicit, non-conceptualizing recognition".

Nonconscious Perception Falls Under Cognition We have also seen earlier in this text research that establishes our processing of information from the environment without being consciously aware of it. As there has been a large amount of this already discussed in the book (see Sect. 2.3), I will say no more about this now except to note that this is a kind of cognition.

Explicit Memory Falls Under Cognition The act of consciously recalling a remembered fact, whether it is semantic (naming) or episodic (i.e., what you had for dinner last night), both require cognitive processes involved in the retrieval of that information. This is both obvious and non-contentious. (Although it is more complex than is initially recognized; see the entry below "Conceptualization falls under cognition".)

Computation Falls Under Cognition This would primarily include mathematical computations. Unlike both conscious and nonconscious perception, computation is untethered from the proximate sensory inputs from the environment and is thus perhaps the most easily categorized as cognition.

Conceptualization Falls Under Cognition This is an uncontentious claim, and so doesn't require any illustration or argument. A few things though should be noted. Semantic memory, which is the naming of things (a part of explicit/nonprocedural system) might in some instances fall under this category of conceptualization, *but not necessarily*. We can see this if we examine the other part of nonprocedural memory—that is, episodic memory. The latter would not necessarily be part of the conceptualization processes, and in those instances when it does not, it would implicate the category of semantic. For example, one could imagine a very young child remembering having had the strawberry ice cream the night before *without realizing* that that particular thing called "strawberry ice cream" fell under the semantic category of "ice cream". These larger, semantic categorizations take quite a while in developing children, and the episodic memory functions would not yet have that semantic category to refer to. Thus, for a long time the token strawberry ice cream would be conflated with a type; or more precisely, there would not yet be a type. Only once vanilla, chocolate, etc., were introduced would the full-fledged type "ice cream" come into being. Then the child would have developed the larger type category "ice cream" and would now see her specific strawberry ice cream as a token of the larger type. Her world had enlarged. (Thank God!)

The category of semantic memory then must be seen as being in conjunction with the episodic; the development of terms and categories in both would be in some degree a lock-step process. This, in turn, makes the semantic category one that is *sometimes* functioning in the manner of traditional conceptualization, but at other times functioning only on a token level where a particular name (e.g. this strawberry ice cream) is linked only to a particular instance. The term is not yet functioning as a type/genus. And, thus, the semantic memory system is not yet functioning as the locale of conceptualization.[13]

What Falls Under "Non-cognitive"? Now that we have listed what falls under cognition, what might be left over that we could put under the category of "non-cognitive"? Quite pointedly, I am genuinely baffled what that might be. It seems to be the null set.

We are now at an impasse. For clearly the claim that perception is the domain of the non-cognitive can no longer be thought to be true; for what kind of processing can be called non-cognitive? In what way, *very specifically*, can one rope off the process of bringing in visual information *in a way that is absolutely non-cognitive*—a way that *does not involve any of the processes listed above?* For even the baby, the dog, or ChapGPT are all doing some version of cognition, and even though we don't want to say it is conceptualization, it would be completely inaccurate to claim any of those processes as non-cognitive. Given this, it becomes very hard to imagine what the term might refer to. How would one *take in information* in a non-cognitive way e.g., in a way that is *not* taking in information? It seems like a contradiction in terms.

And, so, it seems obvious that a large part of the fallacy has always been to equate "cognitive" with "conceptual". More to the point, when theorists were referring to "cognitive" they were using that term as being co-extensive with "conceptual". But there are many good arguments to think of cognition and conceptualization as *not* co-extensive. First of all, we've shown that the category of cognitive includes the conceptual, but that there are many other members of that class e.g., implicit memory, nonconscious perception, procedural processes, probabilistic processes, and sometimes explicit semantic memory. Concomitantly, we've shown that there are ways of processing information from the external world that do not involve conceptualization. But, in these instances, nevertheless, cognition must still be said to be happening.

---

[13] It seems to me that Locke was aware of this fact. See his discussion of general terms in Book I, "Chapter II: No Innate Principles in the Mind" in his *Essay Concerning Human Understanding* (Uzgalis, 2007).

Conclusion

This, then, diminishes the role of concepts, making it merely one of several ways of going from low-level to high-level. To flag up our winnings, this must be admitted: The dog and the baby go from low-level features to the high-level of "food" or "bottle" without ever having to consciously employ concepts. Further research will show the extent of probabilistic, non-conceptual, and non-hierarchical mental process. But, in fact, we don't have to look very far into the future to see the power of Large Learning Models (LLM) like ChatGPT, which have shown this to be true: We know without "knowing". And it's not only ChatGPT and babies and dogs: We do, too. All the time.[14]

In addition, given what we now know about the science behind gist perception, the problem of the role of concepts becomes even more vexing. For this research brings along with it not only the elevation of low-level features, but also the corresponding diminution of the role of high-level ones with their putatively governing concepts. For if we take seriously all the enumerated points in the first paragraph of this chapter, let alone recent developments in LLMs, we are forced to revise the customary role that concepts have in many theories of perception. And as we do so, we then must also revise our notion of the role of both high-level features and the concepts that putatively govern them.

I use the word "putatively" for a reason. Although I have, in the earlier sections of the book, conformed to the standard view that it is concepts that govern high-level features, I think we can now see that that is not entirely accurate. Given the role of implicit recognition, of procedural memory and Bayesian probabilistic categorization, as well as the fact that both semantic and episodic memory do not necessarily employ hierarchical (e.g., general terms to specific terms) reference, and the fact that the data structures and organizing principles of LLM's are concept-free, we can think of the processing of high-level features differently.

In short, we have several different ways of categorizing gist perception into larger categories. Some of these are conceptual and some are not. Thus, concepts have suffered a two-strike hit. In other words, if low-level features are more central to perception than high-level ones, then concepts too are demoted, and we instead see concepts in the revised role of short-cut, *mnemonic devices meant to speed up perception* through the convenient categorization function of general terms. *But even in that short-cut role, not all of the processes are going to be conceptual.* Thus, in a double-whammy, concepts are demoted.

---

[14] Aristotle, when arguing against Plato's notion of innate knowledge in the act of remembering, points to this fact in "Book I" of his *Posterior Analytics* where he states, "Recognition of a truth may in some cases contain as factors both previous knowledge and also knowledge acquired simultaneously with that recognition….we should perhaps say that in a manner he knew, in a manner not" (Aristotle, 2000, 1).

The obvious should be noted at this point: The term "perception", as used by vision scientists, incorporates both low-level and high-level features, even though the semantic nature of high-level features employ (sometimes) concepts. Thus, clearly, this use of the term "perception" by visual scientists folds cognition into perception, at least as those two latter terms are used within philosophy. This might pose a problem for us unless we make an effort to keep these two uses—i.e., the one in vision science and the one in philosophy—in mind. I might though give a spoiler alert, as I'm going to ultimately argue later in this chapter and in subsequent chapters that we ought to frame this issue more along the definitional lines provided by the vision scientists.

3.2.3.2 Concepts' Role in Representation

We can now address the theory of perception that is currently most accepted: Representationalism. While there are many different versions of representationalism,[15] the following are the broad strokes that are generally agreed upon. The general notion of a "mental representation" sees visual perception relating to the world through the representation of that world. These representations are said to have contents and those contents are said to be abstract objects such as properties or more commonly propositions. Thus, those contents are said to postulate the existence of semantically evaluable properties (e.g., reference, truth-conditions, etc.), thereby giving those contents objective truth conditions. The objectively evaluable semantic contents then stand to the world in terms of reference. This referential relation is central to the theory, which is intentionalist at heart; thus, representational theories are often referred to as "intentional". In this, they *refer*—they are *about* something.

The role of semantic properties and propositions is central in these intentional theories. Because such intentional mental states stand in relation to things in the world and explain the latter in terms of their semantic properties, it is therefore the common view that those semantic properties are in the form of the abstract objects of propositions. This, of course, comes from Frege. Furthermore, to access a proposition is to be consciously aware of those semantic properties. If one says, in English, "The snow is white", the predicate "snow" (what Frege called "the sense") is mapped onto the subject "snow" (what Frege called "the object"). When the sentence is said, for example, in German, "Der Schnee ist weiss", it not only has the same grammatical structure i.e., subject and predicate, but more importantly both sentences refer to the same proposition. And this proposition is an abstract object. Thus, the perceptual fact in the world that snow is white is understood through the

[15] Though representationalism takes many forms, including the content view, the object view, as well as a strong and weak versions of intentionalism, I will not address those as they are beyond the scope of this book and will instead in this subsection only refer to the fact the conceptual versus nonconceptual debate happens here as well. See Sect. 3.2.3.1. Also see (Dretske, 1981; Tye, 2006).

mental representation that has its semantic content in terms of the form of a proposition. This makes experiences analogous to conscious thoughts. And those thoughts are evaluable objectively as their truth conditions are available for intersubjective examination. To give another example: I have the mental representation that John F. Kennedy is dead, and that representation is a representation of the states of affairs of his mortality. Thus, I can examine the semantic properties of my representation and determine its truth or falsity in terms of the facts in the world.

The idea that experiences have accuracy and truth conditions is central to this view; the relationship between experiences and the mental representations of them is a direct relationship.[16] But though perception is a relation with the world, it should not be construed in such a way that the object of awareness is the representation. Rather, the representation is the way in which one accesses the world—the object of awareness is still the world. This allows a construal of the relational connection between the perceiver and world to be more closely on the lines of direct perception, avoiding a Humean sort of indirect perception or the similar pitfalls found in sense datum theories. Thus, there is no "veil" between the perceiver and the world. In this it purports to uphold a direct realism according to which we directly perceive the objects of the world, rather than some intermediary objects such as the (now abandoned) early twentieth century theory of sense-data or the more recent qualia theories.

But what of gist perception in this story? And what might be the worries we would have on a representationalist theory?

We want our philosophical theories to be explanatorily adequate and also metaphysically acceptable, but we also want them to be consonant with science. And you want an explanation of perception such that the phenomenological experience is taken into consideration at the same time as the pragmatic demands for truth conditions are satisfied. In this, there is always the debate between mind-dependence and mind-independence reality: Are the phenomenological experiences my (solipsistic) ones or do they have a direct causal relation to things in the world? While we ought to agree with David Marr that our perspective on the world is ineluctably private (my own body is at a very particular angle to the world and in a very particular time and space), and that our systems of perception map out a much more objective picture of the world, we also have to admit that accounting for phenomenology has always been a tripping stone.

There seem to be two problems here: (1) Much of the phenomenological experience—the 'what it's like' of experience—is not captured by the linguistic descriptions seen in a semantic content. As many have noted, visual experience is much finer grained than are semantic terms. We can, for example, recognize many more shades of color than we have names for, and that is only a very mundane example of the fine-grained nature of perceptual experience; thus, the difficulty of reducing

[16] Of course, we are sometimes mislead, which is the source of the debate over hallucinations and illusions. Much has been written about this and it is not pertinent to the issues of this book. My interests are in accounting for how correct perception is accomplished and how that is able to play a role in the communication between the artist and the viewer.

qualitative states to naturalistic explanations i.e. the famous "explanatory gap". This makes it fallacious to claim that semantic accounts, governed by conceptual terms, can properly account for experience. (2) Much of perceptual experience is not experienced consciously. The phenomenon of nonconscious perception is now viewed with increasing consensus, and there is additionally a growing sentiment that the role it plays it greater than what was initially thought.

Both of these points are difficult to account for within the broad stroke view of representationalism. Thus, the difficulties of the limitations imposed by (1) conceptual frameworks and (2) conscious awareness, are such that representationalism struggles when accounting for phenomenological experience. Representationalism limits visual experience (and the belief that is attendant upon that) to the realm of conscious, linguistic awareness embedded in propositions as a result of its semantic framework. Others, too, of course have noted this, and various representationalists have attempted to obviate those problems, arguing for nonconceptual content with representationalism (Tye, 2006; Peacocke, 1998). There have also been those who are argued for a place for nonconscious perception within a representationalist framework, for example David Rosenthal who writes: "Perceiving is not always conscious. Perceiving can occur subliminally, as demonstrated in masked-priming experiments, in which an individual reports not seeing a stimulus that nonetheless affects that individual's subsequent mental processing (Rosenthal, 2010, 374)."

But, in my view, those revisions are attempts at shoe-horning a square peg into a round hole; they seem inconsistent with the underlying demands of a semantic and referential account of perception. Conscious perception and conceptual frameworks seem baked into representational theories of perception.

Thus, it is that the consensus view holds that intentionalist theories maintain the separation between perception and conception. And this not only makes it hard to account for the feedback nature of perceptual experience (as the separation leads us to a feedforward model), but it also leads us down the dead-end assumption that equates cognition with conceptualizing. And that we have seen is not true.

Furthermore, intentionalist theories of representation do not seem to answer what I have devoted this book to: *What is the perceptual experience that is happening at the moment we are standing in front of an artwork?* Because I would insist on the contribution of the subject's point of view in determining the phenomenal character of perceptual experience, I thus emphasize the role of gist in perception. It gives us not only quantitative and referential information on an externalist account, it also gives us the "what it's like" of phenomenal experience. And it allows for a framework that moves us from low-level features to high-level features (1) without the necessity of invoking concepts in that transition (e.g. implicit memory, Bayesian probability categorization, etc.), and (2) it allows for mistakes and fallibility in the employment of those concepts.

This is in contradistinction to the role usually assumed for concepts, which are thought to be those inviolate heralds of rational thought: The presence of concepts, in that story, gave us the conceit of our own reign as rational creatures. Armed with them, we could see The General Category (writ large), we could rise to the higher genus, we could manipulate invisible and abstract thoughts. We were, gratefully, something other than mere beasts.

But I would argue that the romance of concepts is something we can give up. Thus, my emphasis on biology and the de-emphasis on traditional notions like concepts. Though concepts *can* serve as markers for bounded sets, which in turn allow us to quickly see and to maneuver, there are pitfalls in that. Our mistakes happen when we have flagged up the wrong set—e.g., applied the incorrect concept or semantic category: We mistake the field for a lake.

But remember, the mistakes were not made with the low-level features: we saw the glittering of the horizontal lines; that much was true. Our mistake was thinking the glittering horizontal lines represented a liquid (e.g., lake) when in fact they represented recently harvested wheat on a hot, hazy day (e.g., field). Thus, we can be wrong about high-level features in ways we cannot be wrong about low-level ones. Because of this, concepts themselves cannot be thought of the guarantors of either truth or rationality. They must be demoted. In consequence, low-level features end up being more important and more reliable than the markers provided by concepts.

3.3 Feedforward and Feedback Models in Perception

3.3.1 Feedforward Models

Let us let that though dangle for a moment and note that this usual view regarding the relationship between perception and cognition is highly parasitic on the model that sees information in a feed-forward process: First there is perception and then there's cognition. This view was also present in the early vision research of Hubel and Wiesel (1962), as well as the research of David Marr in the late 1970s and early 1980s. Marr's well-known book *Vision* (2010) (posthumously published in 1982 after his untimely death in 1980 at the age of 35 from cancer), has been extremely influential, particularly his "2½-D Sketch", wherein he presents the view that we do not actually see in three dimension as we do not see all of our surroundings. Marr's ground-breaking work is considered on par with the work of Mary Potter and set the stage for much of vision research. As noted by vision researchers Lee et al. in a 1998 paper, "David Marr's model for the computation of the meaning of images has dominated theory and experimentation for the last 20 years (Lee et al., 1998, 2429)." But the researchers also note, "A central tenet of this model, however, is the decomposition of the visual processing into successive feed-forward steps, into low, intermediate and high level stages, and the belief that visual cortex could likewise be divided into areas occupied with low, intermediate and high level operations (ibid.)."

Marr thought of the world in the way most people do: That objects exist and that our eyes are good at discerning them. In that, the epistemology is realist, externalist, and mind-independent—this is really the way the world is, and our eyes are mechanical devises that discern that. His perspective is thus not a constructionalist one whereby we are, in essence, making it up as we go along; there was no Goodmanian worldmaking here. As he notes in the introduction of his book,

> From a philosophical point of view, the approach that I describe is an extension of what have sometimes been called representational theories of mind. On the whole, it rejects the more recent excursions into the philosophy of perception, with their arguments about sense-data, the molecules of perception, and validity of what the senses tell us; instead this approach looks back to an older view, according to which the senses are for the most part concerned with telling one what is there (Marr, 2010, 6).

But despite his avoidance of the fanciful and a desire to stick to the straight and narrow, so to speak, Marr was not hesitant to emphasize the enormous complexity of vision, and to read him is to recognize a scientist who was able to translate disciplined, rigorous techniques into an imaginatively conceived notion of the larger picture. But this picture was, as Lee et al. noted, within a basic model that was feed-forward with complexity growing at each stage. For example, in his description of the grouping processes involved in perception Marr writes,

> The raw material from which everything starts is the primitive description obtained from the image that we called the raw primal sketch. One initially selects roughly similar elements from it and groups and clusters them together, forming lines, curves, larger blobs, groups, and small patches to the extent allowed by the inherent structure of the image. By doing this again and again, one builds up tokens or primitives at each scale that capture the spatial structure at that scale. Thus if the image was a close-up view of a cat, the raw primal sketch might yield descriptions mostly at the scale of the cat's hairs. At the next level the markings on its coat may appear—which may also be detected directly by intensity changes—and at a yet higher level there is the parallel-stripe structure of these markings (Marr, 2010, 91).

This view is encapsulated in his famous "2½-D sketches", wherein we see from the perspective of our bodies and construct the dimensions onto the objects that surround us. For Marr, the basic structural arrangement that underwrites the larger picture was framed in this sort of way: There is progress, with each stage building toward an ideal representation of the external world—a building that is achieved incrementally through the accrual of component parts that, through their successive rearrangements, become more complex and thus fulfill their goal. This isn't meant to ridicule or even to criticize. In truth, we all have basic presuppositions that govern what we see; and many of those presuppositions are unknown even to ourselves—that is the nature of presuppositions. Those presuppositions then in turn serve as handmaidens to the articulated theory. But in the end, Theory rules, for it tells us how to parse the minutia provided by discrete observations, which then determine how the whole is structured. And Marr's presupposition was that complexity is arrived at in a feed-forward way, and it is Progress.

3.3.2 The Rise of Feedback Models

This feed-forward view of things began to change toward the end of the twentieth century. To quote again the research of Lee et al.,

> The purpose of this paper is to argue first on theoretical grounds that the low level visual computation cannot be completed before high level computations are begun; second, to present and examine neurophysiological evidence that V1 is computing different types of information during the 40–350 ms post-stimulus time period; and thirdly, to interpret these findings as part of a extrastriate/striate feedback loop in which V1 plays a highly specific role much richer than simply carrying out the earliest stages of visual processing (Lee et al., 1998, 2429–2430).

The writers give a specific example of this in the following description:

> To find the boundary of an object, the first set of contrast edges must be discounted and the second set must be enhanced or interpolated. But an object must be segregated from the background and its boundaries defined before one can compute its shape properties…In other words, figure-ground segregation and object recognition are intertwined: they cannot progress in a simple bottom-up serial fashion, but have to happen concurrently and interactively in constant feed-forward and feedback loops that involve the entire hierarchical circuit in the visual system (ibid., 2430).

Note that figure-ground segregation is usually thought of as a bottom-up and low-level process, whereas object recognition is thought of as high-level. Their point here is radical: *the two functions of figure-ground and object recognition are intertwined and not sequential, thus making low-level and high-level intertwined and not sequential…*

They note also that the feedback functions explain well the fact of top-down influences (ibid., 2447) and that similar results have been obtained in computer models:

> Computer scientists have proposed for several years now that boundary detection cannot be successful with sophisticated local edge detectors alone, but is best accomplished interactively with surface reconstruction…and this computation architecture has been explicitly proposed in preliminary forms for the visual cortex in luminance boundary detection and in texture segmentation. Grossberg has also arrived a similar neural network structure… (ibid., 2449).

Dynamic systems such as neural nets are rarely feed-forward (an example of a feed-forward neural net would be some instances of climate change modeling or other modeling where the nodes don't form a loop, these are used to approximate functions). As one of the most fascinating and amazing developments in technology, neural nets have given us everything from robots that learn, to Swedish music streaming services.[17]

More to the point, feedback loops are, I would argue, the essential part of dynamic models that better represent the perceptual processes than do feedforward

[17] This reference is to the monopolistically ubiquitous Spotify, which has all but negated the role of the music critic.

ones that suppose complexity to be the result of accrual; the latter being a kind of throw-back modeling—almost as though it were modeled on the Industrial Revolution's notion of progress e.g., with each step we make progress and, ah, get closer to perfection. Feedback loops, on the other hand, model a more chaotic, multivariate, and less idealized environment.

Many other researchers since then have emphasized the same fact of feed-back loops within a dynamic model. In 2003, researchers Lund et al. note in their research on cortical columns of the visual cortex, "Moreover, feedback connections from extra-striate cortex target the clusters of neurons that provide feedforward projections to the same extra-striate site (Lund et al., 2003, 22)".

To roughly fill in some of the blanks here: The extra-striate is part of the visual cortex and is located next to the striate cortex. This area receives information from the LGN (lateral geniculate nuclei) and transmits this to the visual cortex; hence, an important function in guiding eye movements. While Hubel and Wiesel's initial modeling of the cortical columns (the cylindrical column of neurons that run perpendicular to the cerebral cortex) saw the process not only as feed-forward and progressively more complex, but as also functionally related to the anatomical separation of the columns. Lund et al. have modulated that older view as a result of their research and confirmed both the intercortical circuity, part of which is the result of the fact that "feedback pathways provide surround modulation of receptive fields responses (ibid., 15)". Initially thought to be functionally important, and to function in feed forward ways, recent research of the cortical columns has challenged that view (see Horton & Adams, 2005). The current view is that the system is a dynamic one in which feedback loops are essential. In the words of researchers Shen et al., "Neocortical feedback is critical for attention, prediction, and learning…Our findings argue for multiple cortico-cortical feedback motifs implementing different dynamic non-linear operations" (Shen et al., 2022, 6389).

I will cite just a few more research results supporting the role of feedback processes in perception. The researchers Tatler et al. published work on saliency maps and the effects of both scale and time on visual choices, focusing on the viewing of paintings and images. Noting that, "The activity in low-level feature maps has been proposed to underlie saccade targeting…A predominantly top down selection mechanism may also result in non-random selection of low-level features" (Tatler et al., 2005, 643–644). As brain states are determined by the interactions between various cortical areas, it is the modulation of lower-level processes by higher level processes that have garnered much attention. These top-down influences are, importantly, the result of feedback connections, and in perceptual learning the recall of that information, in turn, modulates the more simple and basic low-level information. This is of course what is involved in attention, which makes the act of attention then tied not only to feedback processes, but to the mutual determinations of high-level and low-level features. Clearly, it takes two to tango.

And in 2007, the researchers Garrido, Kilner, Kiebl, and Friston note "backward connections are necessary for recurrent interactions among level of cortical hierarchies. This is the theoretical cornerstone of most modern theories of perceptual inference and learning" (Garrido et al., 2007, 20961). The trend toward seeing feedback processes as essential in any modeling has continued. In 2015, the researchers Herzog et al. are quite pointed in their opposition to (traditional) feed-forward models in vision:

> To cope with the complexity of vision, most models in neuroscience and computer vision are of hierarchical and feedforward nature. Low-level vision, such as edge and motion detection, is explained by basic low-level neural circuits, whose outputs serve as building blocks for more complex circuits computing higher level features such as shape and entire objects. There is an isomorphism between states of the outer world, neural circuits, and perception, inspired by the positivistic philosophy of the mind. Here, we show that although such an approach is conceptually and mathematically appealing, it fails to explain many phenomena including crowding, visual masking, and non-retinotopic processing (Herzog et al., 2016, 9).

Put bluntly, the feedforward model does not work. Their argument in summed up early in their paper:

> Still, as we will show here, low-level vision is not stereo-typical, hierarchical, feedforward, and retinotopic, but a complex recurrent process in which the neural representations of elements across large parts of the visual field interact. More importantly, our results question the idea of a simple isomorphism between outer world states, basic circuits, and low-level percepts, as well as classical hierarchical and feedforward models (ibid., 11).

It is a frontal assault on Marr's theory and his presuppositions.

They also make the (to me) obvious point that "Low-level feature processing depends on high-level features as much as the other way around" (ibid., 15). These criticisms are also an indictment against the philosophy that underwrites the feed-forward, isomorphic view. The thought is explained more fully in the following:

> Clearly, high-level processing does more than only modulate low-level vision. It seems that the appearance of the whole, i.e., the output of figural processing, determines performance on the constituting elements, similarly to Wertheimer's conclusion that 'the whole determines the appearance of its parts' (Wertheimer, 1923). For the same reasons, our results question the idea of an isomorphic relationship between external world states, basic circuits, and perception. Since everything appears to depend on everything, vision cannot be explained by simple, basic circuits. The output of a neuron sensitive to Vernier offsets seems to depend on highly complex processing including low- and high-level interactions (ibid., 6).

What we have here is the embrace of a system that not only accommodates a chaotic, multivariate, non-hierarchical and non-idealized environment, but one that accepts that that is how we perceive and thus make meaning in the world. Unlike Marr's notion that the world is just how it is and we are awfully good at representing that in a retinotopic and mirror-like absorption of reality, these recent theorists promote another view well stated by the researchers at NYU who have developed mathematical models of vision: "A lot of the things you think you see you're actually making up," said Lai-Sang Young, a mathematician at New York University. "You don't actually see them."[18] In a related paper, wherein she and two other researchers

[18] https://www.quantamagazine.org/a-mathematical-model-unlocks-the-secrets-of-vision-20190821/. See also (Chariker et al., 2016).

also from NYU explain the successful mathematical model of vision that they have developed where the retina delivers sparse data to the LGN (lateral geniculate nuclei) that in turn sends that data to the visual cortex, they explain that it is here—in the visual cortex—where the "heavy lifting" occurs (Chariker et al., 2016). What we think of as seeing happens not with the eyes, but with the visual cortex part of the brain. It is here in this part of the brain where, the "seeing" occurs.

But even this is over-simplified, for the visual cortex is also receiving signals from many other parts of the brain. They note that their model has the following important implication:

> Even in input layers to the visual cortex, the system is less feedforward and more dominated by intracortical signals than previously thought...This study points to the need for paradigm shifts in neuroscience modeling: greater emphasis on population dynamics and, where possible, a move toward data-driven, comprehensive models (ibid., 12368).

This is extremely important. Vision is a whole-brain exercise, which is significantly controlled by feedback and dynamic systems. This cannot be over-emphasized.

3.4 Art and Gist Perception

3.4.1 The Putative Division Between Perception and Cognition

We have established that the inductive identification of high-level features is more likely to be a source of error than the perception of low-level features. We have also established that concepts are useful as a mnemonic device that serves to speed up perceptual processing, but less useful as guarantors of truth. And, importantly, that concepts are not singularly representative of cognition. Cognition is much broader and more diverse. The problem before us at this point is to determine the relationship between perception and cognition.

With this we are brought to the issue of what model to adopt. Not only does all the research I have cited serve to dissuade us of a feed-forward model in visual science, but I will argue in this subchapter that it should also *dissuade us of the traditional philosophical boundary between perception and cognition that is predicated on a feed-forward model.* This is not only because the latter is modeled on a feed-forward process wherein the percepts are first given and then operated on by the cognitive faculties (an important point), but because the scientific research itself persuades us that the philosophical definitions of both perception and cognition lack coherence. We shall see this once we look closely at the processes, but it is again worth noting that science does not adopt the philosophical definitions. What I mean by that is this. As I mentioned in Sect. 3.2.3, the term "perception" in vision science includes both low-level and high-level (semantic) features, the latter of which involves to some degree or other the employment of concepts. Philosophy, on the other hand, generally uses the term "perception" to be mutually exclusive of cognition, which is the locale of concepts. I will argue that we ought to adopt the scientific usage blurring of the distinction between the two. Although I will argue for an inverse categorization: Science sees the processing of semantic high-level features as a kind of perception, whereas I will argue we ought to instead think of perception as a kind of cognition instead of something separate from it.

To work our way to this conclusion, let us review some of the facts. This review will also serve to motivate our preliminary step, which is to negate the traditional boundary between perception and cognition. Since it is the case that vision is processed in both feedforward and feedback ways instead of only a feedforward way, it also must be remembered that the most successful algorithms modeling vision are not only on a feedback model but are also modeled with low-level and high-level as intertwined *instead of sequential* (see Lee et al., 1998). This non-sequential relationship between low and high level features is not obviated by the inductive relationship between them. In other words, induction does not entail a strictly sequential, one-way relationship going from low to high nor a feedforward process. I have presented research that confirms that we go back and forth between low-level and high-level: We edit, we re-check, we go back and re-look at the low-level particularities, we change our minds about semantic categories once we realize we've made a mistake, etc. Much of this is done using Bayesian probability dynamics instead of the more directed and conscious notion of a particular instantiating a concept (e.g., that is an apple), where perception is modeled on a step-by-step basis with increasing complexity. This is nicely described by the researcher diCarlo when he compares two different paradigms that can be used, describing the first one where "each successive visual area serially adds more processing power so as to solve increasingly complex tasks" (DiCarlo et al., 2012, 423). This Marr-like algorithm is analogized with an automobile assembly line and then contrasted with an algorithm that is analogized to a large organizational structure like the U.S. Army: "In this framework, feed-back connections between the different cortical areas are critical to the function of the system. This view has been advocated in part because it is one way to explicitly enable inference about objects in the image from weak or noisy data (e.g., missing or occluded edges) under a hierarchical Bayesian framework" (ibid.).[19]

Thus, while we do process low-level features inductively toward high-level semantic categories, it is also clear that this is not done strictly sequentially in a feedforward manner. In other words, while it is true that we begin gist primarily with low-level features and that they are fundamental, the inductive process between low-level and high-level does not force us into a model that claims we begin at the low-level and proceed one-way and sequentially to the high-level, never to return.

This fact then blocks the following move. If someone were to argue (mistakenly) that low-level features are the constituents of perception while high-level features are the constituents of cognition (as a result of equating low-level features as

[19] See also (Lee & Mumford, 2003; Rust & Stocker, 2010; Colombo & Seriès, 2012; Rescorla, 2015) for a further explanation of Bayesian models, whereby the information garnered from a group of neurons does not yield a specific value but a probability density function.

non-cognitive and non-conceptual,[20] and equating high-level features with semantic general terms and thus as conceptual and synonymous with cognition[21]) they would be thus deprived from making the argument that perception precedes cognition. The reason for this is not only because we no longer have a feedforward model (a very important point) but also because, as we've seen, the category of non-cognitive is null (unless one wants to equate the scarcity of data in the form of electrical impulses that are delivered by the retina to the LGN as "percepts"—an unsatisfying definition!), which then makes such a mistaken approach unable to claim that perception (e.g., low-level), was providing, as it were, the fodder for cognition (e.g., high-level). Nor would they, thus, be able to argue that there was a clear causal or temporal division between the two. That part seems rather obvious, but there is something else to consider.

Let us stop and think for a moment about the model provided by Chariker, Shapley and Young, as well as the concurring research currently being done in psychophysical fields as well as computational ones, some of which I have been citing. In these, we are given a different theoretical option than what we have traditionally had in philosophy. Traditional perception is thought to give us a rich description of the world. It is, in John Searle's words, "an openness to the world". Our eyes see out onto world. And cognition takes those basic inputs and operates on them, forming concepts, abstractions, and reflections. These definitions are standard, and this is the view that has held for hundreds if not thousands of years.

But this is not the story we have from Chariker et al. In that, the eye's perceptual functions are quite weak: there is a *radical sparsity of data that is delivered by the retina to LGN*. This means of course that it is the "matching functions" in the visual cortex that are doing most of the work as the processing goes from the V1, V2, V3, V4 and into the layers of the inferior temporal cortex (IT). But it also means that the retina has an extremely rudimentary role. This original data provided by the retina to the LGN is so sparse that for a visual area approximately one-quarter the size of the moon, there would only be 10 nerve cells connecting the retina to the visual cortex through the LGN, but for every 10 LGN neurons that travel from the retina to the LGN and then to the visual cortex there are about 4000 neurons in just that initial input layer of the visual cortex.[22] This system is largely thus feedback. To quote the researchers in a related journal article regarding the orientation selectivity function of the LGN and the visual cortex:

> What distinguishes our model from feedforward (or modified feedforward) models (e.g., Priebe & Ferster, 2008; Persi, 2008; Rubin et al., 2015; Goris et al., 2015) are the model predictions we have to offer. Predictions such as the sharpening of OS, higher OS in E cells than I cells, smooth orientation maps on fine spatial scales, orientation-tuned gamma synchronization, and diversity of OS and orientation preferences within local populations are

[20] While they might be nonconceptual, they are not non-cognitive.

[21] While high-level features are uncontentiously semantic general terms and thus conceptual, this does not mean that the conceptual is co-extensive with cognition.

[22] https://www.quantamagazine.org/a-mathematical-model-unlocks-the-secrets-of-vision-20190821/

> based on model phenomena that emerge as a result of dynamic cortical interactions. Feedforward models, which downplay dynamic interactions, generally do not enable such predictions… (Chariker et al., 2016, 12381).[23]

It is a feedback and dynamic model that can accurately predict the (empirically proven) phenomena such as the sharpening of the orientation selectivity (OS) of certain neurons, etc., thus forcing the conclusion that the feedforward model is not adequate to the empirical description.

The feedback model, combined with the sparsity of retinal data, succeeds in retiring the feedforward model, but it also gives us a few questions. On the feedforward model where perception was seen as mutually exclusive of cognition there was a story that seemed clear. There were two different things, they were mutually exclusive, they were sequential, and they separated out the animal part of us (i.e., perception) from the human part (i.e., cognition). Animals share in perceptual functions but not in our kind of cognitive ones—reasoning was ours. But if we assent to the facticity of science and agree on feedback models that incorporate dynamic systems wherein the processing of low-level and high-level features are not sequential *but intertwined*, where the retina provides sparse information to the LGN which they (there are two LGNs) convey to the visual cortex—where the "heavy lifting" is done—where there are many, many intra- and inter-cortical communications between various parts of the brain that assist in the "recognizing" of inputted data, and where much cognition is governed by probability, then: Where are we, theoretically speaking? More specifically: *Where*, then, is the "non-cognitive" process, which is central to the very definition of perception?

And even more radically, what is that process called "perception"? We have seen the problems inherent in the general explanans of the term "non-cognitive", but we can underline that inadequacy if we examine the problem from a neurological and physicalist perspective. The puzzle extends as well to those ephemeral things that are putatively thought to be the percepts themselves. *Where* might those percepts be and *what* might they be? If we were to take seriously current scientific research and combine that with the usual, customary definition of perception as a quasi-mechanical "non-cognitive" process and thus as a thing distinct from cognition, we would have the following view: Perception proper must be identified in a restricted manner as only those electrical impulses absorbed by the retina. Now, admittedly, that is a hard nut to swallow. If, on the other hand, we want to identify perception with whole-brain activity as recognition and categorization involve processes from many different parts of the brain, we have not found a "where" at all. That, too, is a problem.

Or, more fairly, it is a problem at least for those searching to divide perception from cognition. Perhaps, then, it is their problem. For to make perception genuinely non-cognitive we (rather, they) would have to separate out the process from any parts of the brain that can be fairly identified as computational, even if leaving aside the more well-known function of conceptual thinking. That will be difficult, and the

---

[23] See also Orban (2012).

difficulty arises in part from a definitional problem. Philosophically, perception has always been intimately entwined with such things as recognition: Perception is an "openness to the world" ala Searle. It gives us the world. And that involves recognizing what it is we see.

But this clearly is a problem. For we traditionally have not just identified perception as non-cognitive, we have also identified it as the seat of recognition. *But if we do that*—i.e., if we were to hue to the traditional notion of perception wherein perception has always been intimately entwined with recognition while also being non-cognitive—we have a problem, because it seems obvious that recognition resides in cognition broadly construed. This is especially true if we remember all the forms of recognition, including implicit, Bayesian probability density functions, and the role of the nonconscious, not to mention the involvement of the subcortical structures and the limbic system. Or, to use another example, think of LLMs. Those machines are able to recognize a pattern and thus predict what the next word ought to be; the machine is "recognizing" the pattern on the basis of probabilistic functions, which are a kind of cognition broadly construed. But where in this process is the perception separated from the cognition?

Tracked slightly differently, if we were to try to articulate *what* perception is in terms of what a percept is—we would have to do that consequent to answering the "where" question. In other words, we would have to first track where the entity we call "percept" is—where it is located in the neurological and physiological process—in order to fully know *what it is*. But to that we have no answer.[24] The difficulty of establishing the terrain of the (perceptual) non-cognitive follows from the explanans of cognition. We saw in Sect. 3.2.3 that cognition was a much broader category than is often thought, as well as realizing that the role of concepts is much less. Again, this realization is even more forced on us today as the result of AI and LLMs. Now, if we think about the putative notion of perception from a physicalist and anatomical perspective, which we've been doing, the impossibility of locating the "where" of perception forces us to conclude the absence of a "what" in perception since we cannot without fear of contradiction define it. This appears to sink the boat completely.

3.4.1.1 Conclusion

Given the broadly construed definition of cognition and the impossibility of articulating the identity and locale of the putative notion of perception, it is thus obviously impossible to ascertain a division between perception and cognition. For we have a notion of cognition that incorporates not only multivariate ways of absorbing information from the external world and manipulating that information, but it also sets aside no domain that we could call non-cognitive, nor—even more—a domain that

[24] A related problem would be the one regarding the point at which would one would want to claim that cognition was happening. What would define that moment? And what part of the brain would be assigned to it? These are serious theoretical and definitional problems.

we could call non-cognitive while simultaneously identifying it with the function of recognition. The conjunctive aspect is absolutely the death-toll.

With this, we give up several troublesome problems that have plagued many philosophers. To again reiterate a primary problem: On the feedforward model that sees perception as mutually exclusive of cognition, it was also very unclear exactly what the constituents of those percepts would be such that we could call them "non-cognitive". There were fights and disagreements, and in fact, the problem regarding how to define perception and cognition and how to determine their causal relationship birthed debates that might have served to provide topics for dissertations, but solutions to problems—alas, no.

But adopting a feedback model conjoined with abandoning the division between perception and cognition doesn't solve all the problems. For example, how do we account for the what-it's-like of phenomenon? And how does that play into the various cognitive processes that we listed in Sect. 3.2.3? To do that, we must look anew.

3.4.2 Gist Sensation as Part of Perception

So, let us look anew and try to formulate an answer to these two questions: (1) How do we account for the "what it's like" of experience? and (2) What role does the "what it's like" play in what we will still putatively call "perception"?[25] I will argue that answers to these two questions are found in an account of sensing. But this new account of sensing does not enter into epistemological explanations in the traditional, first-person (solipsistic) way.

Let us take a step back and remember not only Nagel's definition of first-person sensing i.e., "what it's like", but also Evans' notion of sensation: "…the senses yield non-conceptual information, whereas language embodies conceptual information…". The difficulty of parsing this comment of Evans' is found in our common assumption that perceptual experiences give the semantic categories (conceptual content) embedded in language, which then forces the puzzle of what nonconceptual might be, which in turn leads, as we've seen, to the (other) intractable debates in philosophy.

But if we instead rely on the analysis given in Chap. 2, and thus give a different and, I think, more proper analysis of sensing in terms of initial gist perception, we easily find how we might want to model a picture of the epistemological role of vision within the groundwork of sensing. For perception is, largely, a sensorial experience of low-level features that in turn become constituents of linguistic naming. (See Sect. 3.4.3 for a fuller explication.)

Granting that the "feel" of any gist experience is ineluctably private, we must remember nevertheless that gist is also—and importantly—a registering of both the external environment *as well as* one's personal sentiment regarding the impact of

[25] I will shortly replace this term with a more useful one.

that external environment on one's own body e.g., the affordances. For example, if I have a gist perception of a beach, the data processed conveys both the basic shapes, textures, colors, etc., as well as ascertaining whether the scene is orderly or disorderly, spatially open and safe, etc. I also, in this case, respond to the gist experience of the beach with an "ah, how relaxing". And these forces—*both the quantifiable data from the external world and the more emotional valences of affordances*—are vectors that are funneled directly into one person and one person only. Thus, Reality (read: *my* reality) is given to me and me only and only on my terms. It is my perspective (remember David Marr on this point). And it involves the absorption of facts about the world while *simultaneously* allowing for my emotional valences to that world. The two are inextricably married.

Thus, I know both the beach *and* how I stand to that beach. In under 300 ms.

Interestingly, the nonconceptual quality that Evans spoke of is related to this, and if seen in this light can remove some of the confusions around the term. For as we've seen, much of gist perception is of particularities, what I am calling the p-unitates; a term that incorporates both relational (and thus establishing truth conditions) as well as phenomenal particularities. Experiencing low-level (non-semantic) p-unitates is the singling out of particularities and this central fact shows the process as tethered to the body. But not necessarily to linguistic identification; we can experience low-level features without rising to the level of semantic naming. Recognition is something extra. For perception in general and gist perception in particular is not only foveating and saccadic scanning but is of course the relation between those retinal processes and the many parts of the brain that coordinate in order for both experience and recognition to happen e.g., the subcortical structures and the limbic system: Wants, fears, desires. Thus, perceiving is a bodily process.

But saying it is a bodily process is perhaps not enough—while correct, it is not precise enough. To get a better handle of Evans' notion of nonconceptual and of sensing's role in perception we ought to *start with gist*, look carefully at its constituents, and then track that phenomenal experience in its path of developing linguistic naming.

For it is in *that* process we can see where the traditional philosophical definitions of perception and cognition flounder, for once we see the relatively *seamlessness of gist and its relation to cognition, the impassible divide between nonconceptual sensing and conceptual linguistic practices disappears.*

So, instead of thinking in terms of the bifurcation between perception and cognition (as we have seen that no real divide exists) let us reflect on the division between sensation and naming, emphasizing the sensation aspect. (The naming aspect is addressed in Sect. 3.4.3). For how we get to naming things from sensations is the crucial step. This is where philosophy has often tripped: When philosophers assume the division between subject and object, and start with the object, then that transmission from object to perceiver causes philosophers to ask whether those sensorial experiences are still of the external, material object, or are they now sensorial, subjective, or even mental? This has been the difficulty with most accounts of sensing.

But we need not fall into that trap. We need not agree to the strict subject-object bifurcation. For, although it is fair to designate an "inner" world and an "outer" world, in fact the two are experienced together, *simultaneously*, in gist. There is no division at that moment. We register the external data and simultaneously we register our emotional affordances with that data, the two moving forward in time in a lock-step fashion.

Concomitantly, because our registering of the external world is largely done in terms of the particularities of our environment—the low-level features—some of this is concatenated into the ontological chunks of proprietary objects with their linguistic referents, and some is not. In other words, while objects and scenes are recognized, and we thus (sometimes) categorize the data into semantic types, many of the low-level particulars simply stay as such: As low-level features. For example, I walk outside in the month of April and notice a "spring" feeling; I won't perceive that illumination (i.e., the rays of light that are more concentrated in the spring than in the winter) as an object falling under a concept as I would with an apple, but I will nonetheless at least behaviorally respond with an improved mood. I might, sometimes, even consciously note the light's illumination in the air as marking the beginning of spring: "Ah, looks like spring". But that diffused illumination stays as a low-level feature—it is Heidegger's storm in the chimney without the chimney. The quality—the sensation—of the concentrated rays is what is evident; in other words, *it is the quality itself that is evident:* As I enter the outdoor space, the spatial quality of the outdoor scene seems more open than it had a month before, and this low-level feature, functioning very early in the gist perception, would be conveyed to me. Consciously or not. Think of how the winter seems so closed as though the sky is lower and the air heavier. But in (the aptly named) Spring it is the reverse. I would have the sensation of "spring in the air" as the altered light rays make the air look as though it has more light and volume, but there is no circumscribed, denoted object to which that illumination attaches. It remains diffuse and distributed. And in fact, it is often not even noticed consciously, but rather often nonconsciously. We might very well only notice our affordances: We are cheerier.

How might we model this perceptual experience? Instead of seeing this (seeing) as a two-pair relata i.e., nonconceptual/conceptual, sensing/linguistic naming, or perceiving/cognizing, and viewing those stipulative dipoles in quasi-mechanical terms, let us update our framework and see it instead in terms of a dynamic process that involves both feed-forward and feed-back loops. If we do, the following becomes evident. First, please note that the two-pair relata examples I gave all had the same format: The first member of the pair was inner/internalist and the second member of the pair was outer/externalist. Both kinds of data—internalist and phenomenal as well as externalist and relational– *are both given in the gist moment.* And these two thermometers, so to speak, allow us to formulate our responses to the world in terms of naming and semantic categories.

Thus, we can give an account of how our sensing presents options for the editing of the plethora of data in gist experience. We are confronted with uncountably many different low-level features with each gist perception, the multitude of which would make it impossible to absorb all of them; thus, forcing the fact of editing and saliency maps. The now-edited sensing (we've honed it down to a manageable and relevant sub-set!) can now interact with our process of naming.

Thus, it is clear why sensing is so central to this process. For sensing not only gives us the low-level features, which not only give us options to choose from, but sensing also gives us our (solipsistic) relation to that external world, telling us what to edit out and what to retain. This is the role of affordances. Like Marr, we recognize that we experience the world from a very singular location in space and in time. The world funnels to each one of us in singular and idiosyncratically specific ways. As we interact *and meld with it*, we look, we re-look, we edit, we reevaluate—making the feedback loops central to our cognitive processes. Hence, the importance of modeling perception on feedback loop processes, and not on a quasi-mechanical feedforward view.

Sensing, thus, is that part of perception that registers the low-level features and their affordances within the Marr-like framework that situates our individual body as the center of the spatial and temporal experience. It is the first interface. It *can* (but need not) become a particular kind of cognition with further processing, whether it be an instance of identifying the discrete objects in the world in terms of their membership in general and bounded categories, or in evaluating the affordances of the global scene. But whatever transformations might be made of the input sensing has provided, the main point is that the sensing is the initial nexus and that is given in gist. Gist sensing is the first entry point into the human perceptual awareness of the world, an awareness of how one stands to that world.

But none of this is done from scratch. For this includes not only the emotional framework that we bring to each gist experience that frames our recognition patterns, but also therefore our prior experiences stored in memory. Gist, as whole scene perception, is also whole-brain processing. And each of those are radically individual for each of us.

Starting the analysis of perception/cognition with gist gives us the ability to not only model vision outside of the dogmatic constraints of subject/object, but to also see the biological and social role of perception itself: I experience not only the world but importantly how it relates to me—the relation between it and me, and the impact of it on me. It is also how I find the I in the midst of that. As well as that I in the midst of all of you.

This is why, in my view, it is important to see the sensory particularities—the p-unitates—in their full identity and not as either properties of objects or constituents of semantic content. They are not properties or predicates, and they are not limited to conscious perception. They are importantly composed of low-level features in gist, and they give us the very basic information about both the external environment and how we stand to that environment. This wonderful fine-granularity of the p-unitates gives us the ability to pick out the relevant from the not-relevant from the vast plethora of data as it comes to us in the sandstorm of experience; it allows us not only to survive, but to enjoy: For the what-it's-like is the source of much of our pleasure. What more could we want?

3.4.3 Perception (Aka Visual Experience) as a Kind of Cognition

Now that we have established the identity of the what-it's-like experience and thus sensation's role in the epistemological process of perception, let us now look more closely at the transition from sensing to naming. Some of this was briefly mapped out above, but we still need to take a closer look in order to properly position sensation to cognition. Specifically: How do we take the phenomenological experience of sensing—which is embedded in the phenomenological experience of gist—and convert that into words? For it is in the use of words and naming where we are thought to be paradigmatically employing cognition. If we can model how this transition is accomplished, then we will see how perception relates to cognition.

To do that, and to in turn see more fully the epistemological role of perception, we ought to first offer a fuller definition of cognition, which is possible at this point in our investigation. We have seen in Sect. 3.2.1 that a definition of cognition as synonymous with conceptualization is too narrow a definition, and that the brain processes that we rightly identify as also computational—including inductive Bayesian probability models, procedural processes, and implicit learning,[26] would also all qualify as cognitive processes. We also found that the registering of gist, which includes the functions of sensing low-level features idiosyncratically funneled to the individual (i.e., the what-it's-like), gives both external data and personal affordances. Thus, the registering of sensing within the gist experience is a kind of, well, "registering"—that is, a kind of cognition; the brain is processing that data, which in turn can become constituents of further cognitive processes. The external data (helped by the useful functions of affordances and their emotional valences) *can* rise to the level of recognition along with the associated conceptual processes, but cognition itself still remains a broader category than mere recognition or any conceptualization that might be embedded in the recognition process.

Thus, cognition is a broad category with several different sub-categories. We can now define cognition: *It is any ciphering the brain performs on information.* I use the word "ciphering" instead of "computation" as to avoid the computer analogies, which carry with them the association of both teleological and unerring procedures. As many of our cognitive processes are probabilistic it would be incorrect to think of all cognition as characterized in a teleological, goal-oriented way with conditions for satisfaction embedded in the process. This definition that I'm offering is (much) broader than the usual one, as the latter includes notions of "understanding" and therefore deliberative and reflective consciousness. But as we have seen there are many mental processes that factor into behavioral or performative actions that nevertheless were not deliberate or conscious. Much of the processing of low-level features of course falls within this category. Thus, a definition of cognition that can accommodate such facts.

[26] Nonconscious awareness might very likely also fall under the category of computational, but further research on this is needed.

Let me say a little about the notion of "the subpersonal" within this context. As a notion in psychology, one can find references to it as the psychological phenomenon "below the level of the person",[27] but many writers have noted that what exactly that *is* remains quite unclear. It has been adopted into some philosophical writings, but in philosophy, too, writers have noted that the uses are inconsistent at best (Drayson, 2012). I too have struggled with locating a referent of the term, and so I point this out because there might well be some complaints that my broadly construed definition of cognition would be forced to include the subpersonal. Of that, I cannot say very much, given that (1) there isn't a clear definition of the term, and (2) there are many mental processes of which we would be inclined to call them cognitive (i.e. implicit learning, nonconscious awareness, probabilistic predictions, etc.) and how to carve those off from even a very vague notion of a psychological phenomenon "below the level of the person" is a mystery to me. So that will be left to the wayside.

Let us return to the discussion at large. Given that we have concluded that the category of the non-cognitive is null, and that the definition of cognition now includes a broader set of instances, we can position our putative notion of perception in relation *to that broadly construed notion of cognition* (and not solely to conceptualization), leading us to frame the questions as "What is perception's relation to cognition?"

But we are confronted with the obvious as we also have a difficulty with the definition of perception itself.[28] As we saw in the earlier sections of this chapter, the traditional notion of perception as mutually exclusion of cognition and yet as inextricably linked with recognition was problematic; and if we adhered to a definition that *remained* mutually exclusive of cognition and recognition then we ended up with only the retinal transmission process, which does not do justice to our folk version of perception.

But we don't want to give up entirely on identifying the referent of the term "perception" and thereby claiming it doesn't exist at all, and, so, I am going to use a different term to identify the process to which we want to refer: *Visual experience*. This term satisfies several things. It is uncontentious that we all have things we would call "visual experiences" and that it is a fundamental way of relating to the world and gaining knowledge of it. This term doesn't presuppose recognition (or conceptualization) and it doesn't presuppose a separation from cognition. It also allows us to easily construe a relation to nonconscious, implicit, and inductive, (as well as conceptual) processes. And visual experience is not so married to recognition that is divorced from sensing.

Thus, it is not hard to move from the traditional (and conflicted) notion of perception to a broadened (and more consistent) notion of visual experience, particularly regarding issues of sensing. Since sensing is a part of what is traditionally

[27] https://www.rep.routledge.com/articles/thematic/psychology-personal-and-subpersonal/v-1

[28] I'm tempted to resort to borrowing Prince's phrase: "The artist formerly known as Prince"—"The phenomenon formerly known as perception." A change of terms will shortly be introduced.

defined as perception, and since gist has embedded in it sensation, and since all imaginable kinds of visual experience—including nonconscious or implicit—begin with gist and have sensing as part of them, then we can easily say that sensing is part of visual experience itself.

What we still need though is an account of how we get from visual experience to naming; and how we then ultimately model visual experience's relation to cognition itself. This will help us in our examination of how sensing translates to words and thus the connection between sensing (that elemental part of gist visual experience) and cognition.

It was pointed out in Sect. 3.4.2 that naming relies on the substratum of social habits that formed the linguistic community. This point will be revisited again in the remaining two chapters as it is the cornerstone of the communicative and conversational aspect of art. I again take the example of having a gist experience of a beach, my ability to likewise name that beach is dependent of course on the substratum of language acquisition by others, and my understanding of those linguistic terms. Thus, I identify my ineluctably private, low-level sensorial experience e.g., these shapes, these textures, colors, illuminations, orderliness, etc., with the private, low-level sensorial experiences of others along with my "Ah, how relaxing!" and I match the words used denotatively. "Beach" points over there, "relaxing" points in here. The social fact gets knitted in, too, as others have also felt that. I am fused in that way, too.

Thus, that *emotional/factual symbiosis* tied to the social fact is what gives me both the sensorial what-it's-like of experience and it gives me also an awareness of the facticity of the world through the placement of those sensorial processes in memory and recognition. In addition, it ties me to the linguistic practices of others. And remember, the groundwork of all of this is the fact that I experience that scene sensorially largely because it is so infused with low-level features. This makes gist immeasurably valuable despite its resistance to strict quantification. Thus, the "what-it's-like" of experience is combined with cognitive recognition and social epistemology, making gist the groundwork of experience.

While this might seem obvious and uncontentious, there are problems lurking. As it was clear to Ryle, it is also clear to us that we cannot immediately go from sensing to language. There has always been an inner/outer divide that seemed impassable to many, and Ryle's refusal to adopt the "ghost in the machine" was one of the most laudable moves in philosophy. (See the earlier discussion of Ryle in Sect. 2.2.3.)

But, unlike Ryle, we do not have to see the epistemology in terms of the isolated individual trying to solve this dilemma. His analysis revolved around arguments against vision as an instance of sensation (or sensing), and so he analyzed the problem in terms of the individual employing the faculty of perception in a non-sensation way; but that does not seem quite accurate. There is another way to approach this. Instead, we can take into account the fact, stated previously in this text, that the existence of a named entity assumes a substratum of past sensorial experiences by individuals who, through their shared and cumulative articulations of those individualistic experiences, have formed the consensus found in the name. In other words, the role of social epistemology must be considered.

For example, when I see the sunset and call it by name, I am linking my experience with the experiences of the innumerable people before me who likewise have had a similar sensorial, low-level visual experience and named it also according to convention. The first to do so remains a mystery to us all, but whoever that person was many tens of thousands of years ago, we can be confident that the biological/sensorial what-it's-like was similar to ours today. And our relation to that profoundly anonymous person is tightly linked as we add our sensorial experience into the shared cauldron that we call naming. Hence, the function of social epistemology. (See Chap. 5 for a fuller account of this.)

But perhaps I have made it seem too painless. The actual worry here *is* the journey of leaping from the sanctum of the "inner" to the factual world of the objective "outer". Sensing is often thought to be inner in the same way as is reflection, and, on the other hand, words are used to denote the outer factual world ascertained through cognition. Perception was the rubber band that was meant to tie the two together—it could stretch from the inner world of private sensing to the outer world gained by ascertaining external facts—and to bridge their differences. Thus, perception stood between sensing and cognition. But it stood on the quicksand of contradictions and imprecision. So, we begin again with an account that is the result of looking anew at the situation. Beginning with a broadened notion of cognition and retiring of the problematic notion of perception, we replace that latter with the more useful category of visual experience.

In this, it could be said that this new paradigm reflects Ryle as well, for his insistence that we theorize independently of the traditional perceiver/object dualism is here within this new paradigm as well. This is an important point. Sensing is of particularities, of p-unitates, and those do not fully belong to the object. Because gist perception is a melding of the external with the internal, in that way it does away with the division opposed by Ryle. For remember, he was intent on avoiding the equally objectionable oppositions found, on the one hand, in the platonist and transcendent entities, and, on the other, the subjectivist idea-psychology of Hume, Locke, etc. In other words, Descartes' inner world, which was home to the ghost in the machine, was as objectionable as the veil of perception lauded by the indirect perception theorists of the early British empiricists.[29] Thus, the mind/body dualism of Descartes is to be avoided as much as the solipsism of indirect perception.

We though can do it differently. For while experience begins with sensing it also ends in naming, and so it is logical to begin at global scenes that have at their core both private sensing and prior social naming, and not begin with the object. It is logical to begin with gist.

[29] For an excellent explication of this see Julia Tanney's "Volume Introduction: Gilbert Ryle on Propositions, Propositional Attitudes, and Theoretical Knowledge" (Tanney, 2017).

But to only switch to global scenes is only a partial answer. What is needed in addition is the following: tracking the sensing gist experience as it becomes a public object.

And here is the main point: To do an epistemological analysis of how we come to know the world by starting only with the antecedently existing object e.g., that object that was already named by others and known as a discreet ontological entity before the moment of visual experience—is to misconstrue the relationship. For it is not the case that visual experience gives us the completely objective, neutral world. This is the wrong model. It gives us the world *within a context, already imbued with meaning*. And here is the obligatory nod to Evans, who argued that nothing comes to us intrinsically without objective content.

Thus, we don't begin with the naked object, taken out of context, and then attempt to understand how we predicate. Instead, I argue we should begin with gist sensing and track how our individual sensing is in consort with others' sensing *and* with their naming. This includes how we, as individuals—and especially as artists—add to that meaning. Gist sensing is *what we use to make objects*—it is one of the main ways we ontologically chunk the world. And we do this not only using our own gist experiences and the sensing embedded in them, but we do it in consort with our affordances which measure how we stand to that world, and how we stand with others.

And, thus, that world to which we stand is significantly a social world. Naming is an essential part of that, as that social fact helps frame our ontological chunking, so to speak, and our affordances in reaction to that ontological chunking. In other words, the individual's sensorial experience is completely underwritten *by the prior* existence of the socially shared existence of that same thing, as well as the social shared commitment to the name.

But although naming is social, sensorial experience is not. In that, we are each on our own. But this is not a moment of private, pastoral repose. Sensing, particularly as seen in gist perception, shouldn't be seen through a soft-focused, gentle lens. It sometimes has that connotation in philosophy when it is opposed to the more hard-nosed, empirical, third-party verification process that underlies objective facts.

Instead, and in truth, sensing is rather rough, for it gives us the world as it comes to us in a sandstorm of data. We are inundated with information every second we peer out into world: It is estimated that the optic nerve registers 107–108 bits of data every second (Itti & Koch, 2001). Even though, as has been pointed out previously in this book, we only register a relatively small fraction of that data either consciously or even nonconsciously, it is still an onslaught, for if we were to hazard a guess and say that we are only absorbing 10% of those 107–108 bits, that is still a machine gun of data that is targeting our eyeballs!

Thus, we must edit, and it is our editing of the sandstorm bits that allows us to separate the wheat from the chaff, so to speak. With that editing, we respond not only to the outer and factual world but also to the values and needs of others in that we have conformed to the rules embedded in linguistic practices and thus are recognizing previously delineated objects. For example, I see the chair. This means that I have allowed myself to ontologically chunk the world as others have and to agree that that bit of ontological reality that I have circumscribed off in my visual field is both worth noting, while simultaneously agreeing to conform to the usage of the word "chair" when I thus direct my attention to that chunked ontological bit. This means that I have responded with conformity—with social cohesion—by recognizing *what* to notice and absorb and thus to recognize what *is valued* by others and therefore reiterating (to myself and to future others) what is important.[30]

3.4.4 Art as Reiterated Gist-Like Perception

We can now champion the obvious: Seeing is a kind of knowing. It is knowing both the external world and each other. This epistemic role is very prominent in humans as we are organisms that especially rely on sight, making our visual experience function as perceptual cognition—a cognition that governs our ability to both respond appropriately to the external environment and to socially engage with one another. This marriage of purposes is essential, and it begins in gist. Thus, perceptual cognition begins in gist perception, which provides both an externalist reading of the world while also providing the armature from which we can build empathic relationships as well as more complex social organizations that fundamentally rely on cooperation derived from value consensus. When I see what you see, when I parse reality the way that you parse reality, then you and I can work together and move forward with purpose.

And art is one of the main ways we have of practicing the gist in perceptual cognition. For art allows us to both parse the external world and to understand the value systems of others, even more than ordinary perception. This fact gives us a better hold on the main question motivating this book: *What happens at the moment we are standing in front of the artwork?* For example, when I look at a Turner seascape, I see not just the seascape and recognize that external fact, but I also—and here is the central feature—recognize and experience the hazed, murky quality of the scene which pivots between gentle repose and the disquiet of the unknown. Oceans are like that: They can be the most calming thing, and they can be the most dangerous thing; and the transition from one pole to the other can be unannounced and fatal. So, Turner has given us demure colors mixed with rapid and quickly changing brushstrokes, making the scene somewhat hard to pin down and firmly grasp; thus, alluring while a bit frightening. Now, I see the world through Turner's eyes. And I am now brought closer to Turner himself despite the fact that he is now-dead and therefore unreal, because I have nevertheless been brought close to "a person"—a human, like me, who sees the world from their own Marr-like solipsistic perspective as the world funnels into their gist perception. Though he, like all others, sees differently than do I. We are not the same, Turner and I. But: In this moment, I have tried on another man's shoes, so to speak. I have seen the world through Turner's

[30] See Chap. 5 for a full account as it relates to art.

eyes—through the experience of low-level features in his paintings—and know his gist experience as he translated his own phenomenal gist perception into those low-level features. I, thus, know his affordances. *I know how he stood to the world.*

This is true of art in general (not just visual art) because understanding art is a way of *practicing* our ability (1) to sort the incoming plethora of data into relevant and irrelevant, and (2) to practice empathy (and thus conformity) to the values and experiences of others. While this book focuses exclusively on visual art, let me take a small detour and generalize.

Art is not just a pleasant diversion, a warm bath. Nor is it an attempt to find the beautiful. To quote Wittgenstein in this, "You might think Aesthetics is a science telling us what's beautiful—almost too ridiculous for words. I suppose it ought to include also what sort of coffee tastes well (Barrett, 1966, 11)." Wittgenstein's focus on meaning and how we understand the inductive process of identifying the meaning of an experience or a word is worth a moment's reflection in this context. It is not only that the borders of the extensional members of a set are fuzzy or vague or impossible to delineate i.e., "how do we know if this particular thing falls under the category of 'beautiful'?", or 'where exactly does the category of the beautiful stop?'. Wittgenstein's point was not limited to this sorites problem. There was a bigger point, and it is this. Knowing the meaning of something—e.g., knowing what the term "art" means—is found in knowing *when* it is art; it is a thing cognized as an existential click, a flash, an awareness of the truth of the moment. It is knowledge that has arisen as an inductive experience, moment by moment, in time. It is like this:

I sit and watch the dancers—movement by movement, feet twisting and flexing, shoulders jutting into others' or retreating, alienation and intimacy entwined as identical twins—and suddenly, I am stopped short by Akram Khan's movements. I suddenly am overwhelmed; his movements speak at volumes that are clear and crisp and have earned the name "real". I experience it as art. To take another example, I stand in front of the Velasquez, performing eye saccades that sometimes stop and linger. I see these things: the even and small rhythmic brush strokes (though still controlled and deliberate) sweep down in disarray; the color (that is still subtle) is yet rebelliously out of place; the subject's eye says something I didn't expect—more direct and nervous than I expected. Suddenly, I am brought up short, and feel my stomach tighten and my heart rush. I am frozen by the Velázquez; all else melts away and I am aware of only this. I am in love! I experience the art. And know it *as* art.

Sensing the world is not a passive act, as the traditional empiricists would have it. As I previously stated, it shouldn't be seen through a soft-focused, gentle lens. Instead, sensing is a difficult thing for us as the world hits us in its sandstorm of information, and sorting through the relevant from the irrelevant is something that takes a lifetime of practice. Art helps with that. For art is a bodily/embodied awareness wherein I have merged with something significant, something which is now a part of me and has changed my constituent make-up: I am now the person-who-experienced-Akram Khan—the-person-who-experienced-Velázquez.

This description is not meant to elevate art because it is emotionally uplifting. Although it is for that reason that art is often noted to be of value in people's lives.

While that is true, that is not my point. My point is that art is an important way we practice *how to know* the world, and *how to empathize* with each other. *Thus, the practice of gist perception in art is a central way we have of practicing the cognitive functions of perceptual experience.* It's an important way we have of practicing gist.

Understanding this is key to understanding what happens at the moment we are standing in front of the artwork. And iteration is central to this. Several paragraphs back, I wrote, "And art is one of the main ways we have of practicing the gist in perceptual cognition." I want to emphasize the practice part of that sentence. No other species makes art, and that is vital to remember. Thus, art is the main social way we practice how to navigate the sandstorm of information that swoops in upon us. It is also a central way we practice how to empathize with others, and thus to form social structures. In art, we experience The Other: we parse those aesthetic moments as instances of someone else's viewpoint—we see the world through the scrim of their life—the life of their body. It is of someone else's experience as they experience their own embodied reactions to life as it funnels—perspectivally—to them. For example, I watch Akram Khan's dancing and I see an important slice *of him*—of his "take on the world" (a phrase often used by artists), of his reaction to that world and how he has edited his own plethora of data in such a way as to give us what is important to him; to give us *his* wheat, so to speak. He is telling us—using these low-level features such as a slight twist here or a retreating flinch there, what has meaning for him, and what he wants us to have meaning for us.

Thus, art is an act of persuasion. And so, like playing dress-up with paper dolls, I get to try on his view of things, to look at the world with his attitude in place and to see the world through that prism of his. It is not only what is often patronizingly called "enlarging", it is more importantly the cognitive experience of absorbing new data: Khan's data. His edited sandstorm.

Visual art is the paradigm example of that. And in that, as in dance and the other forms of visual experience, visual art is governed by the mechanics that govern gist. In all gist, we see, re-see, edit, re-combine, build saliency maps, etc., all based significantly on the experience of low-level features. But in visual art this activity is exaggerated because visual art relies very significantly on the perceiver's phenomenal experience of low-level features, which often escape the confines of language and semantic definition. I can see the Turner brushstroke and *understand the meaning* embedding in it, but I do not have a word for that particular brushstroke or the shape it has taken. It is meaning without a distinct referent. It is referent- and semantic-free, but yet it can stop me in my tracks. It is this that makes visual art the perfect vehicle to communicate the underlying mechanics of perceptual cognition that are dependent upon low-level features.

Let us look at this and jump more deeply into the mechanics of gist perception in art and see how the experience of art is an important way we practice perceptual cognition. It is uncontentious to say that the ability to navigate quickly in a "cluttered visual scene has evolutionary significance because it allows the organism to detect quickly possible prey, mates or predators in the visual world (Itti & Koch, 2001, 194)." We want to know, very quickly, about that lion. This ability is governed significantly by global scene vision. Looking at the whole and seeing patterns is the essential first step before zooming in to any details. This is why initial viewing of an artwork is largely made up of "widely dispersed fixations of short gaze duration (less than 300 ms)" (Nodine et al., 1993, 220). We see broadly first. And this takes practice.

Experienced radiologists, for example, are known to apply a global analysis of mammography images in the detection of breast cancer: "…the expertise was considered to arise as a shift from detailed scanning to a holistic-gestalt-like perception (Kundel et al., 20082008 as cited by Pihko et al., 2001, 1)." Looking broadly rather than narrowly has many advantages, particularly in inductively seeing patterns in images taken over time as well as failing to over-look the obvious, which can easily happen with too narrow of a focus, as is demonstrated in instances of attentional blindness. Looking globally and getting the "gist" of the scene is our bulwark against error.

We do this also in art. In fact, it is in visual art where can gain significant practice at this. We take in the whole of the painting or sculpture before we proceed to process the minutiae. This is where the low-level features of composition, shapes, colors, contrast, and line orientation are first registered on the viewer's perceptual facilities; we see the art globally. This helps in our navigation of all the information in front of us, but it is not enough. With neural stimulations of the optic nerve at 107–108 bits/s, this is still an enormous amount of information even in the form of low-level, so the visual system then proceeds in a hierarchical manner by processing the data via saliency maps. The researchers Itti and Koch state that saliency is not only driven in a bottom-up manner, but also that it is computed "in a pre-attentive manner across the entire visual field, most probably in terms of hierarchical centre-surround mechanisms…this saliency-based form of attention is on the order or 25–50 ms per item (Itti & Koch, 2001, 194)." The terminology "center-surround mechanisms" refers to those processes that allow us to detect local spatial contrast, an essential component of low-level features. When determining what the oculomotor system prioritizes as salient, researchers have found that "neuronal responses are…strongly modulated by context" (ibid., 197). In other words, our initial scanning relies on mechanisms such as contrast, so a single red shape will show up in a field of green or a diagonal in a field of straight lines. Much minimal art relies on this mechanical and low-level part of vision, for example.

Therefore, our vision is initiated by a bottom-up oculomotor system governed by relatively uniform standard of saliency that can also become, in both feedforward and feedback loops, interactive with object-recognition processes.[31] As it would be over-whelming for us to absorb the external world in one global sweep, the perceptual cognitive system thus breaks it down into sequential and serial computations, which can be influenced by top-down mechanisms from higher areas, including the

[31] There has been an assumption in most vision science writers and philosophers that the process of object recognition is a top-down one. But there has been some growing research in the last decade that challenges that, and some of that is discussed in the next few pages. Suffice it to say, I would follow in the lines of this more recent research that emphasizes the role of peripheral vision, non-conscious processes, and statistical and inductive tiling of data.

frontal lobes (ibid.). Iteration plays a central role in this. This has obvious evolutionary advantages, as the bottom-up (low-level) processes quickly prune incoming sensory input. This process thereby gives us an interpretative map of the world while also providing a role for decision and top-down cognitive phenomenology.

Editing then begins with saliency maps but this is not carried out by an a priori process or even governed by fair play rules. Rather, it is a winner-take-all method. Choices must be made and these are often made as the result of previous experience, habit, bias, etc. Itti and Koch describe it thus:

> Itti et al. (Itti, Koch, and Niebur 1998 & Itti, Laurent, Koch, Christof 2000) consider a purely bottom-up model, in which spatial competition for saliency is directly modelled after non-classical surround modulation effects. The model uses an iterative spatial competition scheme with early termination. At each iteration, a feature map receives additional inputs from the convolution of itself by a large difference-of-Gaussians filter. The result is a half-wave rectified, a nonlinear process that ensures that the locations losing the competition are entirely eliminated. The net effect of this competitive process is similar to a winner-take-all process with limited inhibitory spread, and allows only a sparse population of locations to remain active. After competition, all feature maps are simply summed to yield the scalar map at the core of the model (ibid., 198–199).

This is a model that accommodates a more complex dynamic system predicated on feedback loops, and one that, like capitalism, eliminates weaker competition, aggregates to scale, and sees success as an easy-to-understand elevator pitch. This is not meant as a joke: We want a story to tell ourselves when we open our eyes and look out into the world, and that story is highly edited, and fairly simple. Competitors don't get to share the stage, and the elevator pitch is always there.

But this pitch of course varies from person to person. It is both the case that we all begin processing the low-level features in fairly uniform ways, and that divergences primarily begin to appear as we travel through time, thus making our later millisecond gist perceptions less similar than the early ones. Differences in perceptual processes and the conclusions derived also occur between different groups of people.

Interestingly, this is seen when one examines the process of looking at art. Researchers have shown that it is true that there are patterns of differences regarding looking at art between those who have had lots of experience looking at art and those who have had little. (Practice makes perfect, here, too.) For example, it was known by the early vision researchers Buswell (1935) and Brandt (1945) that there were important differences in eye-fixation patterns between art-trained and untrained viewers (Nodine, et al., 1993, 219). More recent researchers, who have compared viewing patterns between those have had training in art (either as art historians or as artists) and those who have not, have used RSVP techniques as well as the technological advantages provided by modern science to advance this line of thinking. Among these researchers, Pihko et al. summed up some of these findings in the following:

> Accordingly, Kapoula and Lestocart (2006) suggested that experts scan a larger surface of a painting than do layperson. Vogt (1999) and Vogt and Magnussen (2007) provided further evidence for different viewing strategies of art-trained and untrained subjects by showing

that untrained viewers spend more time on areas with recognizable objects and human features than do artists (Pihko et al., 2011, 2).

The researchers also note that the scanpaths were more similar within the two groups earlier in the viewing than during later viewing periods. They note, "This finding agrees with earlier suggestions that early viewing is guided more by low-level processes before a stronger involvement of individual strategies comes into play" (Tatler et al., 2005 as cited in Pihko, 2011, 8). Thus, the reliance on scanning a larger surface of the painting, while greater in experts, is seen also in non-experts as well. An earlier study by Nodine et al. presented subjects with paintings at a 12 s viewing and eye position was sampled at the rate of 60 times per second, measuring then the fixations clustered on given areas of the painting (Nodine et al., 1993). Definitionally, fixation clusters of less than 300 ms were considered "short gazes" and those over 400 ms were "long gazes". They concluded, "Over a 12-sec fixating period, short gazes covered an average of 38% of the picture area. Long gazes covered only 17% of the picture area" (ibid., 223). In other words, all viewers, regardless of training, had more short gazes which would be processing low-level gist perception (300 ms or under) as opposed to longer gazes when viewing artworks.

While this earlier Nodine et al. study establishes the fact that both groups use gist scanning more than they do focused fixations, it also established differences between the two groups, which later studies have reconfirmed. For example, Zangemeister et al. reiterate how experts rely on global gist particularly regarding abstract works. They state, "Professional and sophisticated subjects scanned abstract images more globally than realistic images" (Zangemeister et al., 1995, 1017). This, like most studies, focuses on contrasting the viewing of representational works to abstract works. The reasons for this are fairly obvious: the studies were meant to examine (a) whether trained viewers responded more positively to abstract works (yes, of course, they do—and (b) how the saccadic viewing patterns of trained versus untrained viewers differ or are similar. The differences are especially evident with abstract art, which was shown to rely more heavily on gist perception of low-level features. In a study from 2018, subjects were shown images of abstract artworks in a masked exposure of 50 ms, and then asked to judge whether the artwork's composition was "stable, orderly, and harmonious" or whether it was "unstable, disorderly, and unharmonious" (Schwabe et al., 2018). Interestingly, these initial judgements regarding the stability and orderliness of the composition were consistent with later (reconfirming judgements) made at 3000 ms. In other words, if someone initially judged a composition to be "orderly" at 50 ms, they also, upon greater reflection and longer exposure, judged it identically. This consistency suggests that subjects were attentive to the low-level features, extracted at 50 ms of their experience, which not only demonstrated the soundness of the initial gist perception, but the epistemic reliability of gist itself. This reinforces the conclusion that what is significantly constitutive of an aesthetic experience is its focus on low-level features.

Importantly, this process of experiencing low-level features in the aesthetic experience is an iterative one, with the viewer repeatedly scanning the surface, looking first here and then there, eye saccades varying between rapid and slow. This cannot be over-emphasized and is the core of how art functions as practice for perceptual cognition. Repeated scanning—embedded in the fact of iterative saliency maps where the "winner takes all"—is fundamental to the sensing experience of aesthetic perception. One glances, looks again, re-looks; eyes scanning the whole, looking for order vs disorder, for contrast, etc., drawing in the whole and determining one's first reaction in terms of affordances and how one stands emotionally to the artwork. Evidence for this kind of iterative process is supported by research that has shown that "When gist information in a painting is deemed to have sufficient interest to an observer, the second stage of aesthetic processing ensues" (Locher, 2015, 78). That second stage begins with repeatedly scanning and determining the still low-level features of "compositional features and organization" (ibid).

Thus, aesthetic perception is a staged process, with experiences of several different temporal events of separate gist scannings, significantly constituted by low-level features. Even more interesting is evidence that experienced art viewers were more willing to modify their initial gist assessment about a work of art than were non-trained viewers, more willing to explore the work's complexities, and more "unwilling to express an affective response in a gist experience" (ibid). In other words, experienced art viewers recognize that perceiving an artwork means being willing to embrace the many stages of the gist perceptual process and to reassess the judgements that are consequent to the initial gist scanning. For contradictions can arise, and it is only in the final summation, when the winner-takes-all, that judgment can be stated. Thus, the process relies on repeated scanning, gazing, and constantly reassessing the sensed experience of low-level features.

This level of pensive re-iterated gist is specific to art and forms the core of its ability to serve as a proving ground for perceptual cognition. Let me, thus, further emphasize this important point regarding the iterative process in perceptual cognition. Repetition is how we store things in memory, how we learn, how we come to belief. And in vision, too, this is central. An earlier stated quote by one of the central vision researchers makes the point thus, "The model uses an iterative spatial competition scheme with early termination. At each iteration, a feature map receives additional inputs from the convolution of itself by a large difference-of-Gaussians filter" (Itti & Koch, 2001, 198–199). Choices and edits must be made, and those choices must be sequential as we experience vision in time. Thus, research has shown the process of vision in general to be iterative, cumulative, and summed.

As it is in art. When we are confronted by a work of art, we are also at that moment confronted by something that we don't know. We must process it, and that processing is sequential, sometimes contradictory, and always iterative. We look, we re-look, we edit out and don't pay attention to certain aspects, we see and see again others, which causes us to compound the effects of those others. We also register affordances and react with horror or with delight, etc. And in the end, we come to our elevator pitch. And, of course, all this is done very quickly. Speed is important here, which again speaks in favor of the important role of global scanning as low-level processing occurs in the range of 25–50 ms per item.

Traditionally, it has been thought that added to this is top-down attention (most probably controlled by higher levels in the brain including the frontal lobes), which takes around 200 ms or more (ibid., 194). But there has been some recent work from researchers at MIT that challenges this governing role of top-down attention. Rosenholtz et al. state that it is often assumed (without proof, as they note) that object binding (e.g., binding low-level features into semantic objects) requires top-down selective attention (Rosenholtz et al., 2012). But after noting all the difficulties that are encountered if we assume this (e.g., requiring object recognition to have conscious awareness is inconsistent with the fact that we often can perform many semi-automatic activities such as driving home from work "seemingly in the near absence of awareness or volitional shifts of attention"; it is also inconsistent with many models of consciousness), they propose something different. Instead of conscious attention, they propose a process of object identification whereby peripheral vision performs statistical tiling "computed over local pooling regions that tile the visual input" (ibid.). This is an inductive process wherein the periphery can discern patterns and determine the plausible and likely object to which the low-level features would bind,[32] as well as an aggregate process analogized to compression in computer science. As there is a significant loss of information in peripheral vision, it is yet surprising and worth noting that visual cognition still successfully relies on this loosy data in the determination of gist perception.

This predictive model of peripheral vision fits nicely into an explanation of how we experience art the moment we are standing in front of the artwork. Our initial gist perception, which has in it instances of scanpaths, relies significantly on peripheral vision, which not only accommodates Helmholtz' notion of nonconscious perception, but accords with an explanation of how we come to have the affordances argued for by Gibson. For example, to return to the Turner, I respond to the painting with the sense that I could be pleasantly enveloped by the atmospheric gaseousness of the water scene (or I respond with unease at my feelings of instability). This is largely gained through peripheral vision.

This reinforces the conclusion that what is significantly constitutive of an aesthetic experience i.e., what happens the moment we are standing in front of an artwork and not our assessment of it at a later date—is its focus on low-level features, and contra-positively its ignoring of higher-level features. To put it in more philosophical jargon, we may say the low-level features that the aesthetic experience is focused on is pre-conceptual. In this, it doesn't necessarily involve the process of attributing properties to objects, it just requires the awareness of such features; it is, thus, relying on the preattentive, on the peripheral, on the probabilistic.

References

Aristotle. (2000). *Posterior analytics*. Infomotions, Inc..

Barrett, C. (1966). *Wittgenstein: Lectures and conversations on aesthetics, psychology and religious belief*. University of California Press.

[32] Note the previous quote in this chapter by the mathematician Lai-Sang Young when she described how we see objects given the sparsity of data provided by the LGN: "A lot of the things you think you see you're actually making up."

Brandt, H. F. (1945). The psychology of seeing. https://psycnet.apa.org/record/1946-00751-000.
Buswell, G. T. (1935). How people look at pictures: A study of the psychology and perception in art. https://psycnet.apa.org/record/1935-05800-000.
Chariker, L., Shapley, R., & Young, L. (2016). Orientation selectivity from very sparse LGN inputs in a comprehensive model of macaque V1 cortex. *Journal of Neuroscience, 36*(49), 12368–12384.
Colombo, M., & Seriès, P. (2012). Bayes in the brain—On Bayesian modelling in neuroscience. *The British Journal for the Philosophy of Science, 63*, 697–723.
DiCarlo, J. J., Zoccolan, D., & Rust, N. C. (2012). How does the brain solve visual object recognition? *Neuron, 73*(3), 415–434.
Doshi, T. (2023). In Praise of Shape Poetry. *Poetry Foundation.* https://www.poetryfoundation.org/poetrymagazine/articles/159210/in-praise-of-shape-poetry#:~:text=the%20goddess%20herself.-,Language%20is%20never%20just%20about%20what%20is%20being%20said%3B%20it,reimagined%20and%20shaped%20into%20being
Drayson, Z. (2012). The uses and abuses of the personal/subpersonal distinction. *Philosophical Perspectives, 26*, 1–18.
Dretske, F. I. (1981). *Knowledge and the flow of information*. MIT Press.
Garrido, M. I., Kilner, J. M., Kiebel, S. J., & Friston, K. J. (2007). Evoked brain responses are generated by feedback loops. *Proceedings of the National Academy of Sciences, 104*(52), 20961–20966.
Goris, R. L., Simoncelli, E. P., & Movshon, J. A. (2015). Origin and function of tuning diversity in macaque visual cortex. *Neuron, 88*(4), 819–831.
Herzog, M. H., Thunell, E., & Ögmen, H. (2016). Putting low-level vision into global context: Why vision cannot be reduced to basic circuits. *Vision Research, 126*, 9–18.
Horton, J. C., & Adams, D. L. (2005). The cortical column: A structure without a function. *Philosophical Transactions of the Royal Society B: Biological Sciences, 360*(1456), 837–862.
Hubel, D. H., & Wiesel, T. N. (1962). Receptive fields, binocular interaction and functional architecture in the cat's visual cortex. *The Journal of Physiology, 160*(1), 106–154.
Itti, L., & Koch, C. (2000). A saliency-based search mechanism for overt and covert shifts of visual attention. *Vision Research, 40*(10–12), 1489–1506.
Itti, L., & Koch, C. (2001). Computational modelling of visual attention. *Nature Reviews Neuroscience, 2*(3), 194–203.
Itti, L., Koch, C., & Niebur, E. (1998). A model of saliency-based visual attention for rapid scene analysis. *IEEE Transactions on Pattern Analysis and Machine Intelligence, 20*(11), 1254–1259.
Kundel, H. L., Nodine, C. F., Krupinski, E. A., & Mello-Thoms, C. (2008). Using gaze-tracking data and mixture distribution analysis to support a holistic model for the detection of cancers on mammograms. *Academic radiology, 15*(7), 881–886.
Lee, T. S., & Mumford, D. (2003). Hierarchical Bayesian inference in the visual cortex. *Journal of the Optical Society of America A, 20*(7), 1434–1448.
Lee, T. S., Mumford, D., Romero, R., & Lamme, V. A. (1998). The role of the primary visual cortex in higher level vision. *Vision Research, 38*(15–16), 2429–2454.
Locher, P. J. (2015). The aesthetic experience with visual art "at first glance". In *Investigations into the phenomenology and the ontology of the work of art: What are artworks and how do we experience them?* (pp. 75–88). Springer.
Lund, J. S., Angelucci, A., & Bressloff, P. C. (2003). Anatomical substrates for functional columns in macaque monkey primary visual cortex. *Cerebral Cortex, 13*(1), 15–24.
Mandelbaum, E. (2018). *Seeing and conceptualizing: Modularity and the shallow contents of perception*. Wiley.
Marr, D. (2010). *Vision: A computational investigation into the human representation and processing of visual information*. MIT Press.
Nodine, C. F., Locher, P. J., & Krupinski, E. A. (1993). The role of formal art training on perception and aesthetic judgment of art compositions. *Leonardo, 26*(3), 219–227.
Orban, G. A. (2012). *Neuronal operations in the visual cortex*. Springer.

Peacocke, C. (1998). Nonconceptual content defended. *Philosophy and Phenomenological Research, 58*(2), 381–388.

Persi, E. (2008). Two Potential Mechanisms of Spatial Attention in Early Visual Areas. *arXiv preprint arXiv:0806*.4161.

Pihko, E., Virtanen, A., Saarinen, V., Pannasch, S., Hirvenkari, L., Tossavainen, T., Haapala, A., & Hari, R. (2011). Experiencing art: The influence of expertise and painting abstraction level. *Frontiers in Human Neuroscience, 5*, 94.

Pinker, S. (2007). *The stuff of thought: Language as a window into human nature*. Penguin.

Priebe, N. J., & Ferster, D. (2008). Inhibition, spike threshold, and stimulus selectivity in primary visual cortex. *Neuron, 57*(4), 482–497.

Rescorla, M. (2015). *The computational theory of mind*. Cambridge University Press.

Rosenholtz, R., Huang, J., & Ehinger, K. A. (2012). Rethinking the role of top-down attention in vision: Effects attributable to a lossy representation in peripheral vision. *Frontiers in Psychology, 3*, 13.

Rosenthal, D. M. (1993). Higher-order thoughts and the appendage theory of consciousness. *Philosophical Psychology, 6*(2), 155–166.

Rosenthal, D. (2010). How to think about mental qualities. *Philosophical Issues, 20*, 368–393.

Rubin, D. B., Van Hooser, S. D., & Miller, K. D. (2015). The stabilized supralinear network: a unifying circuit motif underlying multi-input integration in sensory cortex. *Neuron, 85*(2), 402–417.

Rust, N. C., & Stocker, A. A. (2010). Ambiguity and invariance: Two fundamental challenges for visual processing. *Current Opinion in Neurobiology, 20*(3), 382–388.

Schwabe, K., Menzel, C., Mullin, C., Wagemans, J., & Redies, C. (2018). Gist perception of image composition in abstract artworks. *I-Perception, 9*(3). https://doi.org/10.1177/2041669518780797

Shen, S., Jiang, X., Scala, F., Fu, J., Fahey, P., Kobak, D., Tan, Z., Zhou, N., Reimer, J., Sinz, F. and Tolias, A.S. (2022). Distinct organization of two cortico-cortical feedback pathways. *Nature communications, 13*(1), 6389.

Shottenkirk, D., & Chatterjee, A. (2010). *Those dumb artists! Amnesiacs, artists, and other idiots*. Nova Science Publishers.

Tanney, J. (2017). Volume Introduction: Gilbert Ryle on Propositions, Propositional Attitudes, and Theoretical Knowledge. *Journal for the History of Analytical Philosophy, 5*(5).

Tatler, B. W., Baddeley, R. J., & Gilchrist, I. D. (2005). Visual correlates of fixation selection: Effects of scale and time. *Vision Research, 45*(5), 643–659.

Tye, M. (2006). Nonconceptual content, richness, and fineness of grain. In *Perceptual experience* (pp. 504–530). Oxford University Press.

Uzgalis, W. (2007). *Locke's 'essay concerning human understanding' : A reader's guide*. Bloomsbury Publishing Plc.

Wertheimer, M. (1923). Laws of organization in perceptual forms. First published as Untersuchungen zur Lehre von der Gestalt II. *Psychologische Forschung, 4*, 301–350.

Zangemeister, W. H., Sherman, K., & Stark, L. (1995). Evidence for a global scanpath strategy in viewing abstract compared with realistic images. *Neuropsychologia, 33*(8), 1009–1025.

Chapter 4
Making: "*Seeing Anew*"

Abstract Aesthetic theories are always from the point of the view of the viewer. This chapter does it differently. The aesthetic perspective of the artist relies on the first person account of her world. The well-known teacher Josef Albers wrote that artists are to "see anew" and "to open eyes"—e.g., artists are to see the world *not* in terms of semantic descriptions. "The chair" is really very many, many chairs. It is the most fundamental aspect of learning how to paint and draw. No skill can be imparted until this mental work has been done. And this is emphasizing the experience of the low-level features of visual experience: The illumination, shadows, negative spaces, the unexpected moments of color, etc. The world comes to us as "seemings"; they are, in Evans' words, "resistible". This means that we have the capacity to resist the original gist visual experience and re-enter the process and look again. The p-unitates are a kind of knowledge by acquaintance, given to us in determinate moments, and processed through saliency maps—a roping off of some low-level ontological p-unitates. Artists recombine them in such a way—with affordances—as to make a useful object for human communication.

Keywords Josef Albers · Peripheral vision · Rachel Whiteread · Vincent van Gogh · Property attribution · First person experience · Negative spaces · Nonconceptual · Seemings

4.1 Introduction

We can now turn to the analysis of visual experience from the point of view of the maker of art. I do this from the point of view of the artist, as I am, in addition to being a philosopher and former art critic, also a practicing artist. Receiving my MFA degree in painting before I returned for a PhD in philosophy, I began as a painter and have continued thus. This experience grants me a peculiar bifocal ability of looking at art both from inside the practitioner's perspective and also from the outsider perspective of the critic/philosopher. As a life-long bifocal, I have always

D. Shottenkirk, *Art as Cognition: How Gist Reframes the Aesthetic Experience as Conversation*, Synthese Library 521,
https://doi.org/10.1007/978-3-032-00143-6_4

been struck how odd the disjunct between the two is, and how often that disjunct goes unnoticed. Or perhaps more fairly, the disjunct *is* noticed but the dismay that it causes has silenced those voices that, instead, successfully articulate only the perspective of the perceiver, simply opting not to address the perspective of the maker. Instead and consequently, the perspective of the artist is often imbued with unnecessary romance and existentialist verve, rendering the profoundly *articulating act* of art-making into the sentimental subjectivity of mystery. This I hope to correct. And, of course, it is with the lens of gist that makes this possible.

4.2 Gist and Artists' Training

4.2.1 Seeing Not Only Linguistic Objects

The visual experience the artist has when looking at the world in an aesthetic manner in preparation for making art is an experience that is significantly different from the perspective of the perceiver of art, and it is here where the preattentive, the peripheral, and the probabilistic become preeminent. Those three characteristics become preeminent through the following facts: artists see not only (or even primarily) linguistic objects, artists see negative spaces, and artists "see anew". The subchapter details each of those things in terms of artists' training.

Not only is this experience of the artist not the same as the visual experience of the viewer of art, but it is even less similar to non-aesthetic, task-driven, ordinary visual experience (which, in turn, more closely resembles the visual experience of the viewer of the art than it does the visual experience of the maker of art). Let us briefly look at the non-aesthetic, ordinary visual experience for an instructive point of comparison to the visual experience of the artist.

Ordinary task-directed vision tends to consciously perceive the semantic objects in the world—the cars, the buildings, or animate objects such as humans. This conscious awareness of semantic objects is not, as we've seen, all that the perceptual brain is doing, as it is also absorbing many non-semantic low-level features, but object identification is important and it is pragmatic. For it is in these moments when we are attending to organizational tasks such as, for example, driving to work, that we identify proprietary objects that are ontologically delineated such as one's own car or one's side of the road. This object-identification is the usual story of perception.[1] And, as we've seen, gist is essential in these visual processes—it gives us a rich world of information and affordances very, very quickly, and helps to keep us alive.

[1] But of course this example also leads us to point out, as many researchers have done in the past, that doing such "automated" activities also involves parts of the brain that operate without conscious attention, leading some to think "the car seems to drive itself to my office".

But artists, in the moment of viewing the world in aesthetic terms, are not doing the usual story. Let me clarify that a bit: Artists of course drive cars and negotiate swerving drivers just like the rest of us, but they also spend time just looking at the world *aesthetically*—letting their eyes graze over the scene in *non-pragmatic* ways. Like gleaners combing the already plowed fields, artists attend to low-level features in the world, picking and digging, selecting and saving, until it is of sufficient weight and substance. The "it" in that sentence is the artist's vision—the version of the world that accords with the artist's first-person, sensed experience.

In order to make this clear, let me give a short description of some of the low-level choices made by and artist everyone knows: van Gogh.[2] His "La Berceuse (Woman Rocking a Cradle)" (1889) is a perfect example (Fig. 4.1).

For it is the low-level choices that determine the "signature" of the artist. For van Gogh, the stylistic extremism of his work was the direct result of a very few choices: he selected the low-level features of (1) *brightly-colored* (2) *irregularly-shaped*, and (3) *complementary colors*, thereby allowing contrasts to sit next to each other and the harsh brightness to shock. At the same time, there was negative editing, allowing van Gogh to ignore and edit out the low-level features that would moderate that clash e.g., avoiding subtle colors that would have gently transitioned or shapes that would have softened and melded into the shape next to it. The result is, of course, van Gogh: there is more cadmium yellow than lemon yellow or pale yellow in his

Fig. 4.1 Vincent van Gogh "La Berceuse" (1889). 36 1/2 × 29 in. (92.7 × 73.7 cm). (Metropolitan Museum, NYC)

[2] In order to view this image in color and high resolution, see: https://www.metmuseum.org/art/collection/search/437984

paintings, more bright reds and lime greens, more irregular polygons than symmetrical squares, and more complementary colors (such as red and green) sitting next to each other or sometimes mixed together—a color-wheel choice that others have usually sought to avoid. For to choose is to avoid, and to avoid others' avoidance-choices is to positively choose. He did both.

As do others. This low-level editing is the way artists make sense of their experience; it is a particular way of gaining sufficient data so that the first person experience of the artist can be translated into a third person experience available to the viewer. And not incidentally, this editing often entails saving the detritus—e.g., the things non-artists don't notice and don't care about—for that is an important part of what makes a particular artist's vision unique and expressive. Like gleaners, then, artists come in and take what others do not want.

How does this happen? This is rarely written about as almost all philosophers who have written about aesthetics or the philosophy of art have looked at the problem of how to articulate aesthetics or the philosophy of art from the point of view of the viewer. Perception theory that looks at art has also been explored from the viewpoint of the perceiver. This book, too, so far, has done that in the explication of gist visual experience as it is relevant to art. But for this chapter, I propose to look at it from the point of view of the visual artist. This is an interesting exercise, not only because it is rarely (if ever) done, but also because it will demonstrate the radically essential role of low-level features in visual experience itself that is shown in the artist's particular emphasis on the preattentive, as well as peripheral vision's probabilistic calculus. For artist's aesthetic experience is not one that relies on the ordinary perceptual experience of the linguistically-defined and conceptualized object nor on properties of that object as instances of property attribution. Instead it relies on sorting out the low-level features from the high-level ones, and carefully choosing which low-level features to care about. It is thus the penultimate example of the role of low-level features in seeing, sorting, editing, and ontologically concatenating reality.

This analysis is shown through an examination of how it is that art training deviates from the usual perceptual processes and how those deviations frame the perceptual processes of the art student/future artists. In this, the analysis relies on the well-trod tradition of art schools, for, as is almost always the case, artists attend art schools and receive training. Many of these heuristics have been established for hundreds of years, and so warrant a closer philosophical look in an attempt to understand aesthetics from the point of view of the artist.

As someone who has both been to art schools and taught in them, I will point out a cardinal rule: You are taught to ignore the objects. If you want to paint, for example, a chair, you are told not to stare at the chair and then paint what it is that you *think you see*, but to attend to the nuances: The way the light strikes different parts of the chair and the shadows that result along with the parts that are now not illuminated at all, or to notice the surface textures and—strikingly—the way the color varies much more radically than what you'd assumed. You are now surprised for you *thought* it was a brown chair, but as you look—very carefully—you see yellows and reds in the highlights and purples in the shadows, and these varying colors add up to a greater percentage of the surface than does the brown. In short, you are surprised to find out that what you thought is simply not true. Now you see it differently.

It is the particularities that matter, not the objects. To give another example, art training also teaches the students to be aware of the negative spaces between the legs of that chair and how each leg sits next to that negative space and what the relationship between them is. This too leads to shock as the boundaries of the chair have dissolved, and the data set that is being concatenated includes the spread between both the leg and the empty space next to it. That is the circumscribed area. That is the focus of attention; the new "object". In short, it is not looking at the object as the ontologically chunked object you'd always known. You must begin again.

Furthermore and importantly, each student artist is going to see this differently, especially once these instructions have been given. One student (or professional artist) will, like a quasi-Mondrian, see it geometrically and will select out the low-level features of line orientation and color. Another will see it almost completely in terms of shadows and lights and thus see low-resolution (gist) blobs, and will (amazingly!) start with the shadows, slowly constructing the positive spaces. Another will see it in terms of spatial categories such as open or closed and thus be more like Cezanne in the three-dimensional translation of the three basic shapes of square cube, round ball, and triangular pyramid.

To champion the obvious: All of these are low-level choices. And all of them are made in the absence of considering the linguistically-defined objects. Thus, the chair must go, if the painting of the chair is to survive. And I didn't even address color choices. Those too are momentous and highly, highly edited. No two students will choose the same palette even when sitting in front of the same still-life set-up. Thus, to once again champion the obvious: Everyone edits the low-level features differently.

But there is a central point here and it is worth reiterating: The linguistically-defined object must be ignored in order for the particularities to be seen. The chair—"the chair"—is really so, so many different chairs. Our philosophical notion of general terms has obliterated this fact so well known by art students. The abstract concept that governs the general term (on some metaphysically accounts) is not only, as Hume complained, eviscerated of all identifying details, it is evaporated into the ether of vagueness. Artists retrieve that now emptied concept and give it, once again, life.

Thus, the perspective of artists—as they have an aesthetic perceptual experience of the world—is one that relies not on the perceptions of high-level, semantic features e.g., on the linguistically-defined and conceptualized object. Thus, it is a break with the usual view wherein the high-level semantic perceptual features are defined in terms of the application of governing concepts (e.g., it is an "apple") and it is the latter's function vis-à-vis general terms that gives us the ability to identify objects as F of x. But the practice of art disagrees. Contrary to the usual philosophical view, the perspective of artists relies on the sensed experience of low-level features of visual experience. It is the particularities of sensed visual experience that rule and not the vaporous concept. To know the world is to know the sensed particularities.

To further differentiate the view of the artist from the non-aesthetic, task-driven, ordinary perception it is necessary to point out that this special emphasis on low-level visual experience is also not an instance of property attribution (as it is in ordinary visual tasks). In the artists' viewpoint, a property—for example, the blob of yellow-green in the middle of each of the flowers in the van Gogh—is not experienced as a thing necessarily belonging to the object it sits on (e.g., the sunflower). In contradistinction, the ordinary perception of a "yellow shirt", for example, *is* experienced as the yellow being a property that belongs to the shirt. This is, in fact, what property attribution means. This is how adjectives function, and what their ontological status is. (It is a bit slavish…) But in art the yellow is instead the vehicle that forcefully pulls both the artist and the subsequent viewer into the intensity of the green-yellow itself. The color is the vehicle for the affordances. It is what takes us to the embodied reaction. Thus, the green-yellow blob is now promoted and is itself an ontological entity; it does not reside as a part of the object flower, but as a phenomenological fact in itself. Therefore, I *experience the yellow*. That makes this aesthetic process a focus on visual experience as sensing, which is not of conceptual objects.

Ontology becomes more crucial in this framework because artists are making the world as they perceive the world. Van Gogh made a world-according-to-van-gogh, and this is what all artists do. The editing and selecting gives a prismed view of reality that carries with it a range of affordances. I am limiting the discussion here, as in the rest of the book, to visual artists, but I suspect that this claim holds true for all artists. Creating is not just making "somethin' purty"; it is not instantiating the beautiful. Rather, it is delineating the low-level edits in such a way as to construct new objects through the act of circumscribing those newly elected low-level features such that they concatenate into a new object. That is my definition of art. Let me give an example.

Rachel Whiteread: An artist known for her uncanny sculptures, a good example of which is the 1991 piece entitled "Untitled (Mattress)".[3] In this sculpture, what is being pointed to isn't normally what we'd already be calling a linguistically-defined object e.g., a named thing, antecedently defined (Fig. 4.2).

Let me describe it. It is a large plaster shape and the plaster has four holes in it at approximately the corners of the rectangle. And though they do not go all the way through the plaster they are clearly reminiscent of the four corners of a bedpost, but something feels a bit off. The size is almost right for a mattress, and it is indented slightly, thus seemingly showing the presence of its former owners in the echo of their most vulnerable and intimate moments.

But this is not quite right as the absence being point to is more philosophical than one might think upon first glance. The plaster sculpture, seemingly pointing the

[3] For a high resolution and color image, see: https://www.luhringaugustine.com/artists/rachel-whiteread#tab:slideshow;slide:28;enlarge:true

Fig. 4.2 *Untitled (Black Bed)*, 1991. Fiberglass and rubber 12 × 54 × 74 in. (30 × 137 × 188 cm). (© Rachel Whiteread; Courtesy of the artist, Luhring Augustine, New York, and Gagosian Gallery. Photo: Lamay Photo)

familiar and named e.g., a mattress, is really pointing to something else instead. This is where the educational function of museum's plaques is useful and it in part reads: "…plaster casts of the space beneath an ordinary double bed, with the four round holes demarcating the space once occupied by the bed's legs…inviting us to see what is not there or to notice details that are normally hidden."

The art object is pointing to what we don't normally name and isolate for view: the space under the mattress. That is what the plaster cast is of. This is a thing we wouldn't normally have noticed and named. But now we are looking at that and thinking of what that means: Whiteread has edited reality for us by pointing to things not normally ontologically delineated. By identifying a swatch of reality not usually set aside as a set of data (for who sees the negative space underneath their bed as a positive object?), she has uniquely seen it as a member set of particulars, she has drawn a line around that set of particulars and allowed us to see it also. Furthermore, she is telling us *to value* those particulars; to value this *new* object.

Thus, our world now has an additional constituent entity that it did not have before—the formerly un-named and not thought-of: The space under the bed. That is, now, an entity—it is one thing, united into an named existent. It has been circumscribed off from contiguous bits of information and pointed to and named as one unit. Furthermore, and importantly, we are told *how* to view it: More forlorn than even the un-made bed itself, this space records, too, the former inhabitants but records them as ghosts whose physical impact on negative space echoes silently in their absence. The affordances are clear: we are hesitant to get too close to this object that carries with it the sad fragrances that can lurk in an abandoned bed.[4]

Epistemically, what is happening here is that the artist has taken her perceptual experiences of the world along with their associated subjectively proprietary

[4] See my review of Felix Gonzalez for a similar artwork. In Review: "Felix Gonzalez-Torres" MoMA Project Room, NYC. *Art in America*. Vol. 80 (November 1992), p. 132.

qualities, what we call "affordances"—those felt experiences that we have when physically confronted with another being or entity that tells us how to respond and what to do—and has converted those experiences into an object, into art. Thus, the experience of the artist is made into the artwork, and the artwork thus encodes and transmits the artist's perceptual experience.

This demonstrates the importance of low-level features in the editing process. Whiteread's low-level feature choices, like van Gogh's, are few in number; they are (1) negative space, (2) the impact on shapes made by other now-invisible objects, and (3) a muted and monochromatic pallet. The latter allows the other two features to dominate. (If you think back to the Aude Oliva's images contrasting the gist of a city scene with an interior space, Whiteread is the low-resolution version of that gist perception.) That is the core of what is called an artist's "practice"—it is an artist's take on the world, the artist's awareness of the world—and it is stored as a procedure, a practice. It is not at all like semantic naming, and is quite probably stored as a memory in the basil ganglia memory system, making art more like riding a bike than like naming species of trees. It is thus that the making of art is one of the central ways that we parse reality and is thus an important constituent part of our cognitive systems.

4.2.2 Seeing Negative Space

This ignoring of linguistically-defined objects in favor of their constituent low-level features extends to the common practice of teaching students how to see negative space. A few paragraphs back I had given an example of students drawing shadows and then constructing the objects as an act of negative identification e.g., the objects were where the shadows were not. Seeing negative spaces is a related kind of seeing. In this, instead of looking at the object and then drawing only that, students are taught to look *in between* the objects (instead of the shadows of the objects); to see the space between them in order to identity *the character* of that space e.g., is it deep, dark, or lightly infused with bits of light, does it serve to push the rest of the scene apart or to pull it together, are we afraid of that space or drawn to it, etc. This must be considered because the space contributes as well to the overall unified character of the scene.

Consequently, the artist must attend similarly to each, what I would call "moment" across that scene if the drawing is to capture what it is in the scene that is of interest for the artist. This is paying attention to every single perceptual particularity, whether that particularity is tied to a linguistic object or not. In this, the same kind of attention and respect must be given for each moment—each contiguous particularity—across the visual panoply. For the negative space has worth equal to the objects. This is for a few reasons. First of all, objects sit in space, and they inhabit the space in very distinct ways. They are partners. For example, a chair that is sitting close to other chairs sits in a crowded way, and that would be shown not in the drawing of the chair but in the rendering of the space between it and its neighbor chair. That space must be shown as condensed, cramped, tight, and dark. As a drawing student, one learns that those negative spaces, if left unattended, ignored and un-articulated, will stand in the way of a good drawing. One must see them.

Secondly, the negative space has equal worth because it contributes equally to the affordances that are elicited by the scene. For example, if the scene is to be pensive and somewhat depressing (think of an Edward Hopper here), then the negative dark spaces have great importance in creating the response of alienation and loneliness. In order to see this point, let's think of a more specific and well-known example of his: "Nighthawks", which depicts three people sitting inside a glassed-in dinner on the corner of a city street. The outside street is dark, and the interior brightly lit, but after this initial bifurcation what one notices is the negative spaces between the plate-glass window, between each of the people sitting at the bar, between the street's doors and the sidewalk, and the insides of the doorways and the windows. The painting is littered with dark rectangles and dark spaces. All of these very condensed and weighted spaces then serve to separate and push apart the objects, including the people, making thus the objects very still in space; they are locked and cannot move. This in turn immobilizes the scene, rendering it frozen and inaccessible to the viewer who is then mesmerized by the intense alienation. Thus, this affordance is delivered, largely, through the treatment of the negative spaces.

Even though they are not named. This, again, is making the same point: Semantics does not govern our seeing. But let us think for a moment, and perhaps conclude by doubting the abilities of language in general. Is this too radical? Namely, (no pun intended) is language isomorphic with reality? And if not, how much of reality is captured with language?

Remember the basic facts of gist: It is holistic perceptual experience within the first 200–300 ms (milliseconds), and it is different from object recognition. Object recognition takes much longer and requires higher spatial frequency, while we see global holistic scenes very quickly and at much lower frequencies. Low-level features include such things as color and spatial frequency, orientation, hue, texture, contours, motion, volume, and illumination. Furthermore, (and not surprisingly) there is an intimate connection with action. Researchers have found that that when low-level scene information is available neurologically then it is available for behavioral categorization. This makes low-level features useful for inductive pattern-recognition and the quick responses engendered with that kind of thinking. This is different from the slower, linear thinking associated with semantic recognition.

In addition, the contrast with high-level features is not necessarily to the detriment of low-level ones, for though test subjects are able to name their perceptions of high-level features *after* the perceptions themselves occur, and though this is not always the case with low-level features, it is nevertheless not true that this signals their diminished role. For low-level features, which are often without names or semantic category and thus not subject to semantic retrieval, are nevertheless absolutely central for us as they guide our further saccades and because they frame our initial affordances. We know "the gist" of what we are confronted with based significantly on the low-level features. This has its obvious evolutionary advantages, but there is more to their role than that: They tell us not only the very basic facts of the situation, whether or not it is dangerous or desirable (these two things used to be more mutually exclusive in our evolutionary past), and thus whether we stand to it as a thing to move forward toward or a thing to move away from. Thus, the brain exhibits "strong sensitivity to low-level summary statistics…100 ms after stimulus onset confirms that this information is available early in processing" (Groen et al., 2013, 18821). Again, we join the chorus: Gist, as global whole-scene visual experience is significantly the result of processing low-level features.

This, though, I would argue, was always known by artists. Peripheral vision is activated when one looks at the negative spaces. It is true that our eyes are socially trained to identify objects as the central feature of our conscious, attentional visual experience. It is though in the periphery where other more interesting things happen. While only about 3% of our retina foveates, the rest is busy with blurrier but computationally robust activity. Dangerous movements, attractive people, abrupt changes in light, all get computed and assigned probable values. For example, an abrupt movement on the sidewalk next to you might initially motivate a worried and somewhat fearful shrinking from the potential opponent, but the milliseconds that follow allow a small child to burst forth next to you and your body resumes its mass and muscles relax. The perceptual brain had seen, evaluated wrongly, then revised, and processed forward all probably within less than a second. And none of that was done with naming.

An interesting fact emerges from this biology. Training one's eyes to see the negative space also trains one's perceptual brain to loosen its hold on the foveating, attentional, and linguistic entities and to allow the roaming that is possible with the peripheral and probabilistic. This, in turn, allows the artist to see patterns unseen by the more restrictive foveating processes and to imagine corresponding worlds that live beyond the borders of depiction. Dreamy, unfocused daydreaming takes on a different psychological hue…

Photography, by the way, claims the territory of reality, but its presentation is in fact as false as false can be: The photograph is entirely in focus with no peripheral image at all. This is in fact not at all the way our eyes absorb reality. The photographic image would be accurate if—in reality—we summed all our individual eye saccades that we had in looking at a scene, as then they would be equally focused. But the temporal requirement underwriting the summation would take significantly more time than the typical few seconds time-frame it takes to look at a scene. In the latter real-life example, we are foveating only on that little bit of information that our eyes are directed to. For example, I am currently sitting in a park and when I look in front of me, my eyes fix on the vines growing on the far wall. That is seen clearly and distinctly. But also in the nearer foreground are children and parents looking at plants, and though I was not immediately aware of them as my focus was on the vine, I could see first their shapes and then quickly saccaded to them and saw them. I focused on them; it was a second series of eye saccades. Also, in the even nearer and even blurrier foreground (as it is not focused on) is the fence that I am looking through in order to see the vines on the wall (and now the parents and children). I've stopped and focused on that, too. My eyes shifted and added that now foveated image to the total picture. Even more peripheral are the people sitting on the bench on either side of me, who I know only by vague shapes and am frankly only assuming are people. Those I could be wrong about, more so than the parents and children, for the people next to me are more peripheral. The more peripheral, the more probabilistic. The more we bet.

What all this shows is that looking is, to paraphrase the mathematician and vision modeler Lai-Sang Young, something we pretty much make up as we go along. Photography, by giving it all at once, eliminates the hierarchy inherent in the choices we make as we shift our eyes and accumulate layers of eye saccades. It eliminates a part of our system that shows what it is that we value. And even more importantly, it eliminates our peripheral vision. That is the region where certainty steps aside and possibility intrudes, having been shown the way through probabilistic calculus. It is the world of creativity.

Focusing on negative spaces allows the artist to not only see the scene more fully and thus to take into account the relationships between the positive and the negative shapes, but to be able to pry apart the norms of perception and see the peripheral and the maybes. It opens up the space for play; for that part of the human mind that sees not just rule-following, but sees the unexpected, the new, the unnoticed. Those are the things that are not just at the core of humor, but at the core of interest. Art is that: Drawing one's attention to something that is startling, relevant, and moving; it stops one in one's tracks, and that is something we value.

4.2.3 "Seeing Anew"

What we've been discussing so far is a summary of not only how artists are trained to see, it is a window into a kind of seeing: A kind of seeing not restricted to the identification of linguistic entities. This kind of seeing has been spoken of in past by many different people, but my favorite description is by the German mid-twentieth century artist and teacher, Josef Albers—a well-known German abstract painter who, after the infamous 1933 Nazi *Entartete* exhibit and the resulting Nazi persecution of artists, moved to the U.S. and joined the John Dewey-influenced Black Mountain College in North Carolina. He wrote, in his 1963 *Interaction of Color*:

> In visual perception, a color is almost never seen as it really is
> -as it physically is.
> This fact makes color the most relative medium in art…
> …What counts here—first and last—is not so-called knowledge
> of so-called facts, but vision—seeing.
> Seeing here implies Schauen (as in Weltanschauung) and is coupled
> with fantasy, with imagination *(stanza form in original)* (Albers, 2013, 1–2).

His comments here reiterate the discussion in Sect. 4.2.1 regarding seeing and then painting the colors in the chair; instead of painting what you think the color is, you need to see anew and notice all the colors you didn't semantically link with the chair e.g., the yellows and reds in the highlights and purples in the shadows. For once you have really, really looked with your eyes and not your mind, then you see that these varying colors and you now know that the brown chair isn't really brown after all…

He explains this more fully in a 1970 interview wherein he states,

> I made true the first English sentence that I uttered (better stuttered) on our arrival at Black Mountain College in November 1933 (right after the closing of the Bauhaus art school in Dessau Germany, by the Nazi's) when a student asked me what I was going to teach I said: 'to open eyes'…I have taught—until 10 years ago—for nearly 40 years, that is almost half of my life. And when I think that over—now afterwards, I come to the conclusion, namely that I did not teach arts as such, but philosophy and psychology of art (ibid.).

Learning how to see anew—in Albers' words "to open eyes"—is the most fundamental aspect of learning how to paint and draw. No skill can be imparted until this mental work has been done. For if one is to paint from so-called "reality", one cannot proceed by just painting one object and then moving to the next. To focus sequentially and linearly on *the couch*, and then *the floor*, and *the figure* dooms a painting. Instead, one must focus on the more amorphous spaces in between the named objects—the negative spaces—and see them as continuous with the named objects. These are such things as the qualities of trailing light seen not on an object or in the "light" as a whole but in each discrete bit of illumination—each particularity. It is being aware of the almost infinitesimal changes from one tiny spot to the next. *Seeing* thus in painting and drawing is not recognizing the named objects and replicating them; *it is exactly the opposite of that.* It is more a matter almost of seeing through the objects, so to speak—so that one can get the "gist" of the whole scene; it is going back over each discrete moment in the scene and treating it as a thing in itself, not as merely an appendage to whatever it is that it might be named as part of.

Quite clearly, this is emphasizing the experience of the low-level features of visual experience. It is the particularities of these low-level features that are experienced as sensed entities each in their own (unnamed) right. Each are sensed in their own particularity, and not sensed according to the semantic descriptions that forces an arithmetically mean and thus abbreviated description of things.

Let me give a particular example of this. The famous 1905 Matisse portrait of his wife ("Portrait of Madame Matisse, The Green Line") shows the shadow down the nose looking green.[5] He painted what he saw—the way it seemed to him at the moment—and not because he would have claimed that the green shadow had objective content—the arithmetically mean description of a white woman's nose would be "white", which would be a fact agreed upon by others and claimed as veridical. But seeing as an artist is being receptive to the way things look apart from the way one *expects* them to look e.g. apart from how objects are defined as members of their respective linguistic category. Seeing as an artist is seeing discreet percepts, discreet particularities that are non-object, non-linguistically identified things. And

[5] To view a color and high resolution of this, see: https://www.smk.dk/en/highlight/portrait-of-madame-matisse-the-green-line-1905/

interesting enough, it is *that* experience that is largely what makes reality feel like and thus *be reality*.

Let us leave for a moment Josef Albers' dictates on how to see, and return for a moment to the biological underpinnings of seeing itself and reiterate some of that information; we shall see how interestingly aligned these two sets of instructions are!

It is not only clear that a scene's low-level features contribute to the global properties of gist perception (Greene & Oliva, 2009, 465), but these low-level features are the basic elements (P-unitates) that are constitutive of the mechanical, computational mechanism that our eyes use to absorb the external world e.g., sampled episodes driven by extremely fast saccadic scans. Occurring at 3–4/s, these are also highly edited. This editing is essential to remember as it is absolutely true that we do not process all that is perceptually available (Tatler et al., 2005, 643). How does the editing occur? While there is disagreement among both researchers and philosophers regarding the exact roles of top-down and bottom-features, there is growing evidence that our editing of our visual experience is largely governed by "saliency maps". These maps are, in turn, significantly constituted by low-level features. To again quote the researchers Itti, Koch, and Niebur, "This model consequently represents a complete account of bottom-up saliency and does not require any top-down guidance to shift attention" (Itti et al., 1998, 1254).

Therefore, our eyes edit through a saliency map, which in turn is activated significantly by low-level features. This is how we turn perceptions into significant usable data; how we go from unstructured to structured data. The fact that it is primarily driven in a bottom-up manner (Itti & Koch, 2001, 194) allows it to be underwritten by Bayesian models and contextual clues that are, in the words of Torralba et al. "a natural consequence of the probabilistic framework" (Torralba et al., 2006, 783).

To now return to the issue of how artists look at the world and Albers' notion of "looking anew". Bluntly put, what Albers was preaching and what the scientists describe are not so different after-all. For each moment of looking at the world is an instance of looking anew. Every time we turn our heads and look at a new scene we are forced to edit, to employ a saliency map, to pick and choose in conjunction with our affordances of the scene. The only differences between ordinary non-aesthetic visual experience and aesthetic visual experience is that (1) we are normally focused on the pragmatic issues that revolve around previously-described linguistic objects that are so clearly delineated by the definite article e.g., driving *the car* to *the store*, and (2) we are focused primarily on high-level entities rather than low-level features, and (3) the speed of our perception. For artists have stepped outside that world of pragmatic needs and social conformity and so are able to focus, in slow-motion fashion, the underlying mechanisms of vision: To focus on low-level features. Or, to say it slightly differently, to focus on the world independently of those dominating and miserly definite articles.[6] Thus, "seeing anew" can be perhaps be rephrased as "seeing slowly". This is why it is valuable to look at how artists are trained, for it shows us the following. Artists are paying attention to those underlying low-level processes that underwrite all of our visual experiences.

This is the world where we are cured of our premature certainties.

4.3 Doubting and Reinventing

4.3.1 Periphery's Power of Editing

Peripheral vision evinces substantial editing. Since we are only partially aware of what is in our peripheral vision, it is a place where mistakes are easily made, and so we are often forced to revise and edit. But it is also where many interesting things happen, and our brain has developed strategies that build on this, which is why peripheral visual phenomena radically influences the rest of our vision; it is an essential part of the context of the scene and a significant part of whole-scene gist experience.

This is true not only in real-life perceptual experiences but in artworks as well. What happens on the edges of an artwork is often the most functionally significant part of the piece. Thus, it is a locale to which artists know to pay attention. A wonderful example of this is as follows. As part of her research in her Vision and Art Lab at Harvard, neurobiologist Margaret Livingstone has written, "The elusive quality of the Mona Lisa's smile can be explained by the fact that her smile is almost entirely in low spatial frequencies, and so is seen best by your peripheral vision" (Livingstone, 2000, 1299). This well-known "mysterious smile" gives us the following experience: You can't see the smile very well if you look directly at it, but if you focus on the background, for example, the quality of the smile emerges. In other words, the smile emerges in the periphery when the focus is not the smile. If we look at this from a more technical point of view, we can say that the mechanics of peripheral vision radically influences the perceiver's visual processing of the painting. Da Vinci was surely aware of this.

Scientists, too, have paid attention to peripheral vision, and in a much-cited 2011 paper, the visual researchers Freeman and Simoncelli describe the loss of information in the processing of peripheral perception. While Livingstone is correct about the way peripheral vision can alter the over-all low-resolution gist, there is another side of peripheral vision that is important and that is the fact that we simply don't pick up a lot of that information. Citing an early 1970s researcher in the field, J.Y. Lettvin, who aptly described peripheral (e.g. "sidelong") vision as so distinct

[6] To experience how freeing the world is without definite articles see Henry Green's 1929 novel *Living* (Green, 1993). Green, known as the "writer's writer's writer" gives us the world from a largely first person perspective; it is a remarkable exercise in describing sensations and their interactions with conversations.

that "conventional language used to describe perception cannot easily be used in understanding sidelong vision" (Lettvin, 1976), the researchers describe this loss of peripheral information thus:

> Color metamers occur because information is lost by the cones and cannot be recovered in subsequent stages…Our model closely resembles the early stages of these models, but our inclusion of eccentricity-dependent pooling and the invariance to feature scrambling revealed by the metamericity of our synthetic stimuli seems to be at odds with the goal of object recognition. One potential resolution of this conundrum is that the two forms of invariance arise in distinct parallel pathways. An alternative possibility is that a texture-like representation in the early ventral stream provides a substrate for object representations in later stages. Such a notion was suggested by Lettvin, who hypothesized that "texture, somewhat redefined, is the primitive stuff out of which form is constructed (Freeman & Simoncelli, 2011, 1199–1201).

For clarification's sake, "metamers" refer to color stimuli of different spectral distributions that look the same to observers. Because the retina has only three different types of cone cells (these are the ones that perceive color, whereas the rods perceive black and white) so that each type of cone is responding to all colors' cumulative wavelengths, it thus results in a kind of false equivalency (e.g., invariance) wherein the eye is sometimes unable to discern the actual differences. In other words, sometimes, we aren't able to see. This more often happens in colors that are more neutral. The researchers are noting that something similar happens in peripheral vision wherein we are unable to distinguish differences in information within the peripheral field. In their experiment, three different although very similar images were shown (originally at 200 ms but later at 400 ms) with the central foveating part as identical yet the peripheral information inconsistent from one to the other and distorted. The results showed that test subjects were often unable to discern these peripheral inconsistencies. They note that this seems counter-intuitive, as object recognition seems impaled by this process. This loss of information also leads to the common problem of crowding, e.g., the where we are unable to discern objects or data that are crowded or cluttered together.

Why would our visual system be so riddled with error? Let us look for a moment at the brain anatomy of this. The ventral stream is that pathway that carries information from the visual cortex to the temporal lobe. Formerly called the "what" stream, it is now thought of more subtly as the locale of the sematic process in perception and it works in conjunction with the sensory functions of the dorsal stream that coordinates action. Thus, when we reach for our coffee cup both streams are integrating in that process. The ventral stream has thus always been thought to be the essential function for object recognition. These researchers just mentioned above, though, have shown a more complex story. For they have shown that that information in the ventral stream is radically edited; pooling happens at each successive stage along the ventral pathway and the aggregated information not only succumbs to crowding, but now phenomenon exist statistically but not perceptually. In other words, the data is there, we just can't recognize or know it (ibid.).

It might initially seem that our editing of peripheral information, and its reduction down to fewer elements, allows us to focus on that part of the visual field that is foveated. This is how peripheral vision is often thought of: The stand-in performer for the main actor, waiting in the wings for the circumstances to change. But this is probably not correct. For the ventral stream is processing not only peripheral information but central, foveated information as well. So, problems (if they are to be seen in that negative light) with reduction in ventral stream peripheral information would presumably translate to foveated information as well.

Let us though turn this on its head and view the loss of information in peripheral vision less negatively and instead a bit more optimistically. And, thus, to view peripheral vision a bit more favorably. For what happens in the peripheral is more important than merely playing back-up. This is true for a couple of reasons. First of all, it is in the periphery where we can easily see some of the basic editing mechanisms at play that function in foveated vision as well, and so it is by considering the mechanics of peripheral vision that allow us to model a clearer picture of the whole process of vision itself. For example, one of the most obvious editing mechanisms is crowding, which happens very prominently in the periphery but it also happens in the foveated field just as well; in this, we are often unable to discern distinctions between entities when they are crowded together. Another similarity is the fact that we simply do not process a great deal of the data that is before us, as we saw in the analysis of saliency maps. These similarities make peripheral vision heuristically valuable: Probabilistic calculations rule. But secondly, as Margaret Livingstone points out, peripheral vision is important because it changes how we see the things in the rest of the scene. Low-resolution images function as part of a unit and whole-scene gist experience relies on this. And peripheral vision is an essential component of whole-scene, low-resolution visual experience.

But there is a further point to be made that is consequent to the two arguments (1) the influence of peripheral vision on the rest of the scene, and (2) a reliance on the heuristics of that process. And that is the point quoted a few paragraphs back and made by the early vision researcher Lettvin, "texture, somewhat redefined, is the primitive stuff out of which form is constructed". It is texture perceptions processed in peripheral vision that influence our contextual reading of the scene. For example, if we look directly at the side mirror of a car and peripherally are seeing the reflective surface of the rest of the car, that reflective, shiny texture is influential in identifying the side mirror as a car's side mirror. We don't then have to saccade and look repeatedly at the whole car. A little bit of peripheral data is enough to draw the conclusion. The probabilistic calculus suffices. This reliance on texture as the basic building block also thereby establishes proof that crowding is consistent with representation based on local texture statistics. In other words, we initially garner the data of the world through the processing of texture information.

This is interesting. Texture perception seems to be more essential than commonsense would dictate; we often think of texture as something that is noticed *in addition to* the other perhaps more basic things such as color, shape, or line orientation. The main point here is clear: This is, again, a reliance on low-level features in the construction of visual experience. But the more specific point here that the researchers have noted is that this is particularly striking in peripheral vision.

Where does this get us? The processing that we see in peripheral information, where much is lost and edited out, is true not just in the processing of texture, but in the probabilistic functions of vision in general. This should come as no surprise. We need to decide quickly what counts, and so hunches and bets play important roles. Precision counts for less than what we might perhaps hope.

But maybe not. Maybe we shouldn't hope for precision. And maybe, then, we have reason to not bemoan the frailties of our visual system, but to look more on the bright side and see what it gives us. First of all, let us remember that in this guessing game that we call "seeing", much of what is before us is not relevant to us; thus, we ignore far more than we consciously take notice of or even nonconsciously absorb. We are *very* picky eaters. Not only are we picky, only noticing those things relevant to our affordances and concerns, but our pickiness is not precise. We have seen how we might mistakenly assign certain low-level features as belonging to the wrong semantic category e.g., we mistake the raccoon for a cat as it runs across the street on a dark night, or a stranger down the street for the friend Joey. These kinds of sensing mistakes are at the core of rationalists' complaints against empiricism, beginning of course with Plato's *Theatetus*, and continued to, well, now. But there is another level of mistakes on the empiricist account that is more interesting to note, for many mistakes happen not just in regard to semantic categories but also in the perception of low-level data, although these mistakes are of a different sort. In this case, the mistakes are in the selecting and editing. Many different low-level features have to be noticed and concatenated together in order to make sense of the world, and that process of rapid sorting involves, as the researchers Freeman and Simoncelli noted, a lot of discarding. And those things discarded are not there to be retrieved. That is why our perceptual faculties rely so heavily on rapid saccadic moves and gist experience: We are testing the waters, re-testing, picking up what we value and discarding what we don't. In other words, we are making sure that we have the basic P-unitates of information.

Thus, editing is central to visual experience. We look, look again, tilting, angling, and looking again. We discard lots of data not because we have inept systems but because we don't need the data.

This brings us back to gist, and its very functional and causal role. The basic gist of the scene is the first moment of editing, and this whole-scene. But what exactly does "whole scene" mean? The term is often employed, but it is worth a moment's reflection to precisely think what it might mean. Now that we've had a slight detour into peripheral vision again, we might want to think of whole-scene perception not as foveating widely, but as peripheral vision seen instantly. Think again of Livingstone's point regarding the Mona Lisa. Because seeing the smile happens before foveating happens, it is here, in those first few hundred milliseconds, where editing cuts its first slices—is it spatially open or closed e.g., indoor or outdoor?, it is threatening or not?, is it a seductive smile or not?, etc.

Artists not only notice the peripheral, as that is often where relevant and fun things happen, but the practice of looking at the world with the emphasis on low-level features and low-resolution images is also a practice of slowly, slowly looking and repeatedly editing. We see this not only in perceptual experience but in the creation of the artwork itself. To make art necessitates a practice of not only editing the artwork itself—crafting and correcting, changing and altering, giving up often that part loved best in order to make the whole work—but it also necessitates considering the perceptual processes of the perceiver: What will draw the eye first, what will that first eye contact lead to, how does that activate the page and how does the process dovetail with the intended meaning of the piece? Graphic designers have a very simple version of this process whereby they identify the very first thing in the image that will draw the viewer's eye, and then the next thing. As the message has to be simple in graphic design, there cannot be too many layers in the visual experience; one or two places to look is more the norm. Painters, though, generally have more complex and multivariant intended targets in the artwork with the motive of keeping the perceiver's eye constantly moving. Termed "activating the space" the intention is to not give a succinct and quickly devoured message as in commercial work, but to draw in the viewer for an extended time. This replicates the ordinary visual practice of looking, relooking, and looking again even more closely at the world itself. In this, the power of editing becomes apparent.

4.3.2 Relying on the Nonconceptual

Here is an amusing side thought. What gives us the motivation to edit is also what gives us the motivation to not edit. What I mean by that is this: We are motivated to edit in order to get it right, but we are often also motivated to *not* edit because we are apt to think that we have *already* gotten it right. In other words, as C.S. Peirce pointed out, we are hesitant to re-examine our cognitive processes because we are loath to risk abandoning the happy state of belief in the service of seeking truth, as that truth-seeking move would force us to be in a state of anxiety-producing doubt (Pierce, 1877). As Peirce argues, the consequences are that we often don't really care about getting it right as much as we care about being happy, as we are evolutionarily molded to prefer blinkered, blissful belief over the harsh glare of wide-eyed and stressful comprehension. But this particular evolutionarily-selected trait of believing without re-examining the evidence is not always the preferred choice we make. Sometimes we doubt and then bravely proceed to examine again. It is, in fact, what philosophy tries to train us to do.

There is yet another version of this when it comes to artists, whose ability to re-examine and "see anew" is an essential part of their cognitive perceptual experience. For that aesthetic perceptual experience is often linked to unsettling revisions of reality than are the more task driven and socially relevant experiences of our everyday life. While we in our everyday lives might be unwilling to reexamine habitual ways of looking at reality (e.g., is my belief in the superiority of my own country really well-founded?, is my rejection of scientific consensus regarding climate change a rational point of view?, etc.) when it potentially undermines our social relations, this is not especially the case with artists' aesthetic perceptual experiences. This holds true despite the fact that artists' revisions of reality may challenge a social norm as in the case with topics dealing with sexuality or wider politics, but it may also leave social rules untouched as is the case with purely formalist works. The point here is a deeper one than ruffling the feathers of the powers to be. In order to see that, let us switch from focusing on subject matter choices to focusing on epistemic procedures.

For the artist, the experience of viewing the world aesthetically—seeing anew—necessarily involves a reiterated process whereby the cognitive awareness stays in the realm of particularities and/or whole-scene gist. This is in radical counter-distinction from ordinary task-driven perception where property attribution dominates. The aesthetic perceptual process of the artist is a playful, imaginative process of looking past the objects and into the scene in the instance of whole-scene gist, which allows for not only an awareness of the low-level features, but of an iteration of varying potential ways of looking at the scene. For example, in the moment of initial gist-looking artists are trained to shuffle through competing versions of the scene: How does it seem? Scary, or enticing, or frightening? What lines are prominent—the straight ones, or the angled ones, or the crooked ones? What colors do I pull out and emphasize and what do I diminish? What affordances are relevant for me, the artist, and what affordances do I want to make available to the viewer? Looking, editing, looking again, and seeing something different is, as we have seen, the core of artists' training and practice. This is especially true in the very first few milliseconds of visual experience where our mind has initially accessed the data and made preliminary assessments.

What kind of cognitive state is this initial state that leads the artist to see anew? While it is the case that the first gist assessments give us the framework to proceed with subsequent processing, it is also the case that the character of these initial sensings are fungible. What I am calling the "fungible" quality of the initial whole-scene gist perceptions is a quality that also allows us to make the case that they are of a particular kind of perceptual cognition. How might we characterize those initial perceptions?

An examination of the cognitive role of this sensory glancing and re-glancing has been given attention in recent philosophical literature in Gareth Evans' posthumous *Varieties of Reference*. I will reiterate a few of the points discussed earlier in the book with regard to Evans and then address his work within the framework of the artist's use of the nonconceptual. For all perceivers, artists and non-artists alike, our first sensing/glance of something gives us a certain kind of epistemic access, which he called "seemings". These seemings are sensory and provisional; it is the way something first "seems" to us. But Evans also pointed out that while something might, initially, *seem* to be a certain way to us, we can be mistaken about that initial sensing information and our later sensings can diverge. The fact of this provisional sensing of our environment lead Evans to argue that this is a distinct kind of cognitive function—a kind that gives us nonconceptual information.

The category of the nonconceptual was much debated in the 1980s, 1990s, and early 2000s. While it is not relevant to rehash those debates, it is relevant to outline Evans' main points and to then relate those to the concerns at hand regarding the use of nonconceptual cognitions by the artist. For Evans, the distinction between the kind of (potentially fallible) information given in a sensing, nonconceptual experience and the kind of information given in a (subsequent) judgement was thus explained by him as follows:

> The informational states which a subject acquires through perception are *non-conceptual*, or *non-conceptualized*. Judgements based upon such states necessarily involve conceptualization: in moving from a perceptual experience to a judgement about the world (usually expressible in some verbal form), one will be exercising basic conceptual skills. But this formulation (in terms of moving from an experience to a judgement) must not be allowed to obscure the general picture. Although the subject's judgements are *based upon* his experience (i.e. upon the unconceptualized information available to him), his judgements are not *about* the information state. The process of conceptualization of judgement takes the subject from his being in one kind of informational state (with a content of a certain kind, namely, non-conceptual content) to his being in another kind of cognitive state (with a content of a different kind, namely, conceptual content) (Evans, 1982, 227).

In order to understand the category of nonconceptual, it is important to stop for a moment and think about how he is defining the more general notion of content and its relationship to concepts and, thus, the conceptual. The content of a thought is conceptual; in this way, it is a type (as opposed to a token). Therefore, content is a type when it is constitutive of judgements, which are implemented through the use of concepts: "Thus, if a subject can be credited with the thought that *a* is *F*, then he must have the conceptual resources for entertaining the thought that *a* is *G*, for every property of being *G* of which he has a conception. This is the condition that I call 'The Generality Constraint'" (Evans, 1982, 104). For Evans, this governs the flexible combinatorics of subjects and predicates and explains our common-sense notion that a concept can be reapplied in different contexts within different sentences; if *a* is F, then *a* can also be B.

But content would not be a type in the information-states that are constituted by the nonconceptual, for while Evans emphasizes our ability to uncouple and recouple predicates and subjects, he is adamant in his focus on the atomistic, nonconceptual constituents of perception:

> Philosophers have been prepared to attach considerable theoretical importance to recognitional capacities in their accounts of what it is for a subject to have an idea of a property, or kind, of particulars, while relegating recognitional capacities for particulars to a theoretically insignificant position in their account of what it is for a subject to have an Idea of a particular; it is the aim of the present chapter to redress the balance (Evans, 1982, 270).

To have an "Idea of a particular" is distinct from the common practice of "granting recognitional capacities" in the account of what it is "to have an idea of a property"; and the former has been relatively unattended to by philosophers. Focusing on the atomistic particular, (which is in contradistinction to the notion of a property of an object, as that is property attribution), is to see that visual experience includes sensing and is radically distinct from thinking of a property as an instance of property attribution. Thus, it is, as I have argued earlier in this book, a mistake to think of visual perception as only of linguistic, conceptual objects. Instead, the initial sensed gist is an essential component to visual cognition, as is also probabilistic peripheral information.

Evans, too, recognized this. For Evans, the broader notion of nonconceptual thus refers not only to the epistemological process of sensing but thus also derivatively to those ontological entities that are perceived non-semantically. Thus, when we ask the question *what it is that we are sensing*, we now see that sensing gives us a lower ontological level of elemental percepts, a level that is unnamed. As an ontological level that is non-semantic, it is, thus, nonconceptual.

This is the ontological level to which Josef Albers was referring when he directed his students to "see anew"—they were to look below the linguistic level to the constituent parts as they could more easily be brought into a new focus. The reason for this should be philosophically evident. A "chair" will always be just a chair with the lexicographer's label constricting our access, forcing us to experience the chair as an instance of that general term, and thus not allowing that particularly perceptual experience to deviate from those norms. But if we, for example, shift our gaze to whole scene gist (by squinting), the chair is now part of a larger low-resolution scene and so enters into potentially more diverse relations with its neighbors in the scene and thus taking on different characteristics than the more rigidly defined "chair". It is, perhaps, now just a square blob sitting next to an aggressive vertical of the doorway, or it has converted itself into a series of vertical lines that are the spindles on the chair's back. It has morphed. And is, thus, no longer an *x* of F. Or, to give another example, if we instead of accessing the scene through low-resolution whole scene gist, we alternately shift our gaze to the particularities of the low-level features of the chair e.g., its texture on the arms, the color distortions on the seat, the fracturing of the surfaces as the light breaks up the shape, – we see an ontology of the lowest and most nominalist kind. Either way—viewing it macroscopically as part of a low-resolution whole scene or viewing it microscopically in its particularities, we are opening up avenues to sensorially experience the non-linguistic environment.

To relate this visual experience from Albers' point of view to Evans' perspective we need only shift from a narrative of specific sensed experiences to the birds-eye view that characterizes them. For Evans, the initial interface with the environment is one of sensed experience, and it is the constituent parts of perceptual experience that are not perceived conceptually but instead nonconceptually. This is Evans' insight. To rephrase it slightly, the initial nominalist entities are sensed; it is a *bodily* absorption of the data. It is a first person experience and is thus peculiar to the individual—my sensing is ineluctably mine—and it is not identical to the third person semantically-named object. Thus, my sensing of the chair is not the same as the lexicographer's codified linguistic object. Thus, these epistemological events—these "seemings"—give us, as individuals, the first glimpse of our environment. But the seemings are not writ in stone.

And here is the fun part. It is a portal, so to speak, that can be renegotiated and re-experienced. We can edit the seemings. For Evans, to reiterate, seemings are the initial sensory experience of the environment, but it is important to remember that this experience can also present us with objects as well as with low-level features. Though Evans' views did not incorporate any analysis of whole scene gist as that research had not yet been done at the time Evans was writing, much of it confirms that scientific research. For his object-recognition analysis began not with semantic accounts but instead with seemings, and these are not dissimilar to low-level features. And like low-level features, seemings bring with them certain associative meanings and affordances. This, too, opens up motivational possibilities for the editing. A description of this is as follows.

The nonconceptual character of the seemings imbues them with certain characteristics as neither objects nor their constituent parts enter "without objective content". Everything, for Evans, enters with some degree of content because we perceive things within a schema. (Remember the 4000 neurons at the first layer of the V_1 and thus the matching functions of the visual cortex.) This is important. A quote by Robert Stalnaker that I referenced in Sect. 2.2.2 bears repeating on this:

> Evans regards it as important to identify information-bearing states of perceptual systems with states of seeming since he is anxious to avoid the traditional epistemologist's picture according to which the subject receives, through the perceptual systems, sensory data that is 'intrinsically without objective content', but which forms the basis for inferences about the world that causes them (Stalnaker, 1998, 352).

The "traditional" epistemologist would include the Humean empiricist as well as the logical positivist who posits the neutral given and the realist who views our reception of the external reality as direct, completely mind-independent, and thus "objective". Evans is arguing against a view of the world that sees information coming to us neutrally, shorn of meaning, where the objects of perception arrive at the doorstep of our perceptual faculties in a state that guarantees their objectivity. In that story, these purely objective entities can be used in conjunction with our operations of logical inference and thus we can be guaranteed satisfactory truth conditions for knowledge-acquisition.

Alas, it is not so. Instead, they carry content with them, as they are part of an already established schema. But the saving grace is the "resistible" quality of the seemings: We can alter, edit and re-evaluate. It is a version of Peircean doubt. When things seem not quite right to us, we can "resist" the seemings and look anew.

Visual researchers' perspective on the process is not very dissimilar from Evans and from Albers. To reiterate the view from the perspective of visual researchers, low-level features are often thought of as nonconceptual in contradistinction to the high-level and semantic features, which are conceptual. The *activity* of perceptually absorbing low-level features is what they call "preattentive" awareness. It is in contrast to "attention" which involves semantic and conceptual analysis, and this definition of preattentive as excluding conceptual analysis seems so close to Evans' notion of the nonconceptual as to be indistinguishable. Thus, treating the two perspective e.g., Evans' and the visual researchers', as one perspective, we can make the following observations. What I am calling the elemental p-unitates, which are the particular nominalist constituents of the low-level features, is also isomorphic with the characterization of the phenomenon as seen by both Evans and the visual researchers: The preattentive is the nonconceptual, and it is that initial interface with the non-semantic particularities of the external world in the form of p-unitates that gives us a sensory awareness of the world. And that is of a distinct cognitive category.

That cognitive awareness is a bodily awareness, which is initially in the form of whole-scene gist and is not governed by concepts.

We can now discuss the artists and their use of the nonconceptual. If we take Evans' views, the research provided by vision scientists, and the notion of p-unitates, we can easily understand how artists view the world aesthetically. Firstly, it is clear that artists' ability to see anew is dependent on being able to untangle the linguistic object from the particularities that constitute that thing; it is dependent on staying in the moment of sensing and allowing the sensed particulars to have flexible and fungible rearrangements. It is a special kind of cognitive disciple that emphasizes a "recognitional capacities for particulars". Secondly, it is in radical distinction from the perceptual procedure that views visual experience solely in terms of identifying an object as an instance of a concept. Instead, we have the state of the preattentive, the nonconceptual. Thirdly, it is world not governed by the strict matching rules of relating concept to instance. Instead, it is a world where flexible interpretations are constantly in play, where seemings are resistible, revisable, and editable. And this is where the real happiness lies. The state of artistic play.

4.3.3 Freedom and Creativity

The idea of imaginative play has been thoroughly discussed in philosophy. It is perhaps Kant who first comes to mind for this and so it is thus appropriate to give a representative quote: "For the imagination (as a productive cognitive power) is very mighty when it creates, as it were, another nature out of the material that actual nature gives it…" (Kneller, 2007, 52). And while we can all agree with this noncontentious notion that artists are creating something new out of what was given in nature, our focus here is slightly different. For that Kantian perspective leaves unaddressed several questions. Firstly, it leaves unanswered quite exactly how we might describe the cognitive and brain functions that imagination relies on, and secondly, how we might model imagination's role in cognition itself. Let us try to examine these issues.

Regarding the former e.g., regarding how we might describe cognitive and brain functions that imagination relies on, we have some easy access to that information as fMRI evidence points to the parts of the brain that are activated in imaginative activities. Luckily, science has developed quite a bit over the centuries since Kant, and we now see the brain as a processing unit that is replete with temporal stages involving various and different parts of the brain as they contribute to that temporal process.[7] For example, recent research emphasizes the hippocampus in imagination (Comrie et al., 2022). But the hippocampus is certainly not alone in this as other parts of the brain are also thought to be involved in the process including the

[7] A "processing unit" is not to be misconstrued as identical with the computational model seen in Fodor or more generally in cognitive science.

neocortex, the thalamus, and frontoparietal. But even this is not a complete accounting. The list extends beyond that as the reward centers of the brain are also surely involved and those include the mesolimbic dopamine system composed of the ventral tegumental area and the nucleus accumbens (Lewis et al., 2021).

But clearly, this list of anatomical sections of the brain do not give us a model-theoretic account of what is happening in the moment of imaginative play but instead leads us only to the conclusion that the process of imagination, like visual experience, is a not locatable in one organ or isolated process. This should not surprise. Let us then shift our focus from empirical data and examine instead our second concern, which is imagination's role in cognition itself. In order to do that, we will look at imagination in terms of a phenomenological, first-person model combined with a model that describes brain processes on the lines of neural nets—e.g., weighted data points that are able to develop patterns. The latter, which is the model used for LLMs, can be briefly described as follows. If the inputted data reaches a certain threshold, then that data is passed on to the next node. If not, it is not. This then establishes pathways, which then can be recognized as patterns. Hence, LLMs as pattern recognition machines. And similarly, our brains are also pattern recognition machines; neurons reach thresholds and through dendrites (and their synapses) pass that charge on to the next neuron.

In order to understand how this happens on the ground, as it were, in the moment of looking, I propose the following. We look at the phenomenal experience of a subjective perceptual moment as experienced by the artist, and see how pattern recognition develops. We can then relate that story to the neural net model. I argue that this will tell us something about imagination's role in cognition.

Let us start, then, with that story. The artist is surveying a set-up in the studio that they intend to paint. There is a model and the surrounding stage-like setting. Imagine, if you will, something like a quasi-classical figure scene. The usual way of proceeding with the process is as follows. First, of course, the looking begins. But the artist doesn't start with a focused looking at individual objects or discrete parts; rather, the beginning is a scanning—looking broadly at the arrangement of shapes and colors, patterns, lines. Often squinting is employed with these very beginning glances as that naturally mimics gist low resolution shape relations. If you squint, you will see the basic shapes and the contrasts more clearly in their dominant roles. This simplifies the compositional elements and allows for an accurate placing of the shapes in the geometrical space, which then successfully translates the three dimensional into the two-dimensional. The objects themselves don't need to come into focus at this stage and one wouldn't want them to. It is the relations that matter, for that is what determines the placement and the relative weights. The darks, which weigh more, nestle into the lights, and patterns get established at this early stage. Often this is done is a reiterated way, where squinting is repeatedly performed during the rough sketch in phase.

What I have just described should sound familiar as it is close to what Evans was calling "seemings": How something seems at first glance. And remember, this phase allows for great flexibility and revision. What something seems in the first glance is not necessarily what it seems in the second or third or fourth, etc. It is fluid state of cognitive perceptual experience. What the artist-perceiver is doing is trying out different possibilities that might be found to be satisfactory, running through different options and trying out each one for size; it is a kind of problem solving. In order to do this, different options and different perspectives must be rapidly played through and evaluated as to whether or not they are satisfactory. Thus, the visual process of taking in the phenomenal reality is not exclusively one of the eyes or even the brain—the artist will move around and look and re-look, taking in different perspectives before generally returning the same position vis-à-vis the model and the setting. The phenomenal experience then is of the whole organism, the whole body. The patterns seen then reveal also the invariant features of the scene: What areas stay dark or light, what the major shapes are, what the general tenor is of the light's illumination, what color scheme is dominate, what textures stand out, etc.

It also shows the affordances of the scene. The artist is registering not merely the definitional point of understanding how the scene itself can be "used" or "made use of", but how the artist *stands to that scene*. This is important. It is a relational point. This deeply visceral reaction to the scene is the psychological fact of interaction: Inter—action. How we "stand to something" is a measure of how we meld with that thing—how the interaction between the subject and the object then are concatenated into a new entity that combines both.

For example, when I take account of my affordances to a scene or an object, I am playing through the options with which I might interact with that scene or object—if it is a chair I think of how I might sit down in the chair, or lean on the arm, or touch the leather, etc. If it is a forest scene that faces me as I am standing on the road, I play through the affordances of that scene in my hypothetical thinking of it as I wonder about how I can enter into the forest and what will my relationship be? I play through the options whether it would be nurturing and protective or frightening and dangerous, or exciting and adventurous, etc. Therefore, affordances are not to be understood superficially as how I might react to or make use of the object or scene, but instead how I *absorb* that object or scene into my intimate environmental world. A merging. I accept the other into myself and in that consent to vulnerability, enlarge my own identity. I am, for example, now the person-who-experienced-the-protective-cloak-of-the-inviting-forest. My history has been enlarged, and so has my experience, my memories, and my identity.

But to return to the artist in the studio. Remember: The artist's take on the scene is absolutely individualized; it is that particular artist—that particular person's—perspective and experience. I've generalized the process in describing the over-all process of affordances, but it is important to bring ourselves back into the scene of the studio and the artist's phenomenal process of perceiving the studio model and set-up. And it is that individualized phenomenal experience that is then captured in the translation of that first-person phenomenal experience into the (eventual) third person artwork. One maps onto the other.

Thus, this whole process that results in translating the phenomenal experience into an aesthetic artifact is what we call "artistic imagination". And gist plays a fundamental role in this. The original series of perceptual processes are extremely rapid with the collating of rapid eye saccades and repeated whole-scene gist experiences. Therefore, it is not about objects solely or their linguistic equivalents. It is much broader than that. Thus, the whole-scene experience lies as the foundation of artistic imagination in the visual arts. As the twentieth century painter Paul Klee stated, "Art does not reproduce what we see. It makes us see." It does this both for the artist and the viewer of the art. For the artist, the trained process of "looking anew" and seeing beyond the objects allows for the phenomenal experience just described. For the viewer, it allows for a window into that process.

And herein is the freedom for both. For the viewer, art allows the viewer to see reality from the point of view of someone else, and to also see the too-often hidden layers of the perceptual and phenomenal process of looking itself. Gist becomes more evident. The viewer can access that gist realm of perceptual acuity through the experience of simulating the artist's edited choices. For example, the low-level features of color, texture, shape, illumination, etc. are spelled out, the low resolution relations are evident, and the affordance factors are made psychologically clear. For the artist, it is not a window onto another viewpoint or a deeper access to reality, but a portal into the substructure of seeing itself. When the artist looks broadly and absorbs the whole-scene low-level features and their affordances, they have transported themselves into the reality that is the ground-level of perceptual experience. This, then, also leads to a different portal: The balm of escape. The contemporary artist Yinka Shonibare states it best:

> Being an artist is a form of dreaming in a way because what you do in your art in quite different from the reality of the world. Being an artist is an escape from the terrible things that happen in the world…I like the challenge of facing blankness and creating out of blankness…I like to reflect and think about society generally and express things in a very engaging way but in a very fun way.[8]

To know that one can move into that world and let your imaginative play dance through the possibilities, seeing it one way for a brief part of a second and another way a moment later—finally alighting like a butterfly on the right manifestation of reality—is to freely live outside the confines of life and thus to experience the unadulterated force of life itself; to experience the pure seemings as they manifest themselves to the individual. There is much joy in that direct experience of ground-level reality. As the artist Agnes Martin has stated, "Happiness is being on the beam with life—to feel the pull of life…". Art is that vehicle that gets the artist there.

4.3.4 Re-making Reality

For the artist, this freedom is not only both invaluable and profound, but it also comes as something of a relief. There are, after all, options! Relying on reiterated gist and widely sampling alternative viewings gives not just some abstract notion of freedom, but the permission to access reality orthogonally and radically. There is no

[8] See https://www.bloomberg.com/news/videos/2015-06-22/brilliant-ideas-artist-yinka-shonibare

conformity required; it is not "one way only" as in the linguistic world. In that world, the lexicographer, like a dutiful traffic-cop, tells us which way to go and in truth we are grateful; for social cohesion would be absent without linguistic conformity. But that conformity does not exist in the initial (reiterated) gist perceptual cognition. To again use as our example of the artist surveying the scene and looking at the studio set-up: Choices of color, composition, points of emphasis shown with the use of contrast, choices regarding which parts of the scene are left out, how illumination is exaggerated or diminished, which emotional valences are revealed (tenderness, violence, repose, etc.), how affordances are articulated, etc. The artist takes from the scene many, many provisional and alternative views. And the artist is free to choose. This is very different then when we apply concepts to individual instances.

Concepts constrain, but seemings liberate. This is at the heart of the oft-noted rebelliousness of artists. Let us refer back to Evans for a moment to see how this works.

The world comes to us as seemings, and they are, in Evans' words "resistible". This means that we have the capacity to resist the original gist visual experience and re-enter the process and look again. The artist's perceptual process has this exact process as its mantra: Look, relook, and look again. And, mind you, don't see what they told you you should being seeing! It is thus that artists see it as their *duty* to resist—to look at the world differently. It almost seems like a vow one takes one when one becomes an artist, except for the fact that that interpretation has it exactly backwards; the psychological propensity to *be* rebellious is probably manifested quite early in one's life and the, then, subsequent desire to find artmaking a suitable profession fits neatly into that propensity. But leaving that empirical question aside, of more interest is how the "resistible" feature of Evans' views meshes with the general notion of creativity. How this cashes out is relevant to our concerns.

Making something is creating, of course, but that's a platitude of sorts. We have the image of a free-for-all, of a system without rules, an enticing chaos. But, in truth, the process of creativity is a profoundly precise one. This is often over-looked. For the process of articulating one's own sensory first-person account of experience and converting that into affordances for viewers who can, to whatever degree, experience the artist's seemings, is one that requires precision. It is neither haphazard nor without rules. For in order for the artist's first-person experience to be transduced into a third-person version *that is accessible* to the viewer's perceptual cognition, certain rules have to be followed.

Some of these have been discussed, for example, the rules of composition, colors palettes, and other basic design elements. These are elementary, but other more complicated ones of cultural symbolism and being tuned in to the relevant zeitgeist require an understanding of the interpretative symbol systems that must also be at the artist's fingertips. For example, the artworld zeitgeist of the time in the current 2024 (in NYC) has a few different strands: There is an inward-turning, private-world celebration where images of fantasy figures mingle with doodles, which doesn't surprise as we are still coming out of the covid years where isolation ruled and one's own inner life was the replacement for social interaction. There is also on the scene more political work in response to Black Lives Matter movement or LGBTQ+ rights. Finally, there is the presence in the more blue-chip galleries of large, decorative work somewhat minimalist that easily functions as backgrounds in expensive homes. All of these different (post-modern) zeitgeists exist simultaneously, and if one is to participate in the production of artmaking, one must be sensitive to these readings and possibilities. The reason for this is that these sectors in the artworld function as symbol systems and they are thus interpretative scrims through which the audience (varying though they might be) look at the world. One has to know how one's message will be parsed.

Thus, the artist's experiencing of the seemings is only one of the inputs. The artist must also take into account those interpretative, symbolic scrims that are existing in the culture at that moment. Those "its-in-the-air" zeitgeist phenomena are a very important part of artmaking, which is discussed more thoroughly in the next chapter, as it is a core part of how art functions as conversation. But in order to make clear how that functions, it is essential to now drive home the point that the artist is combining the social zeitgeist with the seemings choices that are made in the studio. To phrase it slightly differently: The artist's choices in the studio—which are an almost uncountable number—are made within the social context. In order for the selecting and compiling of the chosen inputs (that are experienced as seemings) to be transduced to the audience in a successful way, it demands that the artist is rigorous and consistent in compiling those edits. To give a trivial example: Colors such as pale yellows, pastel oranges and whites would not be used in a piece that was being viewed through the framing lens of BLM work, unless those colors were used ironically; this is not an Impressionist moment. Instead, the mood of the moment would demand stronger colors such as cadmium reds, blacks, marine blue, etc.[9]

The obvious point is that all of the low-level choices have to dovetail with the social symbolism embedded in the agreed-upon zeitgeist. Everything comes with meaning, and a precise and rigorous editing process is what allows the first-person experience of the artist to be captured and transduced so that the communicative function regarding the audience is successful. In this way, we now have an answer to our question regarding how the "resistible" feature of Evans' views meshes with the general notion of creativity as well with the notion of communication and conversation. It is as follows.

Creativity is the knitting together of the individual's seemings with the filtering system of the zeitgeist, which, if knitted together properly, serves the purpose of social communication through conversation between the artist and the viewer, as well as between multiple viewers.

But let us approach the question slightly differently. Given that the artist's "resisting" in the moment of the seemings is an essential part of creativity, and given that the role of communication and conversation plays a significant role in the choices that are made, in what way might that resistance actively contribute to our concatenation of reality?

[9] See the works of Wangechi Mutu, Simone Leigh, or Henry Taylor, to name a few.

This ontological aspect of creativity can also be seen through the lens of Evans' views. Specifically, his point regarding the fact that seemings do not come to us without objective content is worth reflecting on. Clearly, he is instead claiming that seemings arrive *with* objective content. What is objective content? To have objective content is to have the object arrive at the doorstep of our perceptual process as an instance of something—it already has meaning. In other words, for him, it is to be part of a schema; a part of meaning-constructs.

Let's contrast that for a moment with its opposite. To arrive without objective content would be to arrive neutrally. In the most ordinary sense of the term, it means that reality would be given to us directly and is thus irrevisable. This view, a cardinal one in strict versions of empiricism, was an attempt to avoid subjective distortions and provide a promised road to empirical facts. But this had its problems. To quote Stalnaker on Evans: "If one succeeded in purging the content of perceptual states of their environmental dependence, what would be left is sensory data, 'intrinsically without objective content'. One would be left with the dreaded myth of the given" (Stalnaker, 1998, 351). The dreaded myth, so felt by members of the logical positivists and so feared by those thereafter, was to have the primary empirical members of the epistemic process be utterly without meaning, bearing no relations to the rest of the world, its things, or our conceptual apparatus.

Our current notion of objective content, is, like many other of our current notions, debated. But while the notion of objective content has varying definitions, it seems fair to say that a relatively uncontentious one would be what the conceptualist argues for: We speak of conceptual content when we assume it is objective content and therefore to have conceptual content is to possess the concepts "that characterize its content" (Schmidt, 2015). Objective contents and objective concepts are thus married; this is what the conceptualist will argue for. It is also frequently the case that those who subscribe to the above also subscribe to the view that all perceptual cognitive processing is conceptual. To see is to exercise concepts, which then pave the road to objective content.

In this, Evans differed. But Evans' notion of it is not the prevailing one. For him, there were different informational states between the subjective moments of "seemings" and the objective linguistic descriptions of the world. But both of these states had content. The former was nonconceptual while the latter was conceptual. But nothing came without content: Even the nonconceptual information comes to us with associated meanings, relations to other things, and thus with official residences in our memory systems. In writing on Evans, Adrian Cussins states, "Thought, even at its most sophisticated and most fundamental, essentially entwines elements of subjectivity with elements of objectivity. Put more strongly the conclusion would be that the objectivity of thought entails the subjectivity of thought" (Cussins, 1999). I would agree that this is a fair assessment of Evans' thought. For Evans, the perceptual experiences are simply an earlier event in the chain of reasoning. "The senses are like witnesses who tell us things that we may accept or reject" (Stalnaker, 1998, 348).

This seems about right. And it also is a resounding confirmation of the role of the artist. What the artist is doing with the editing of seemings is a constructivist concatenation of reality: A roping off of some low-level ontological p-unitates and recombining them in such a way—with affordances—as to make a useful object for human communication. This is the main point: Artists make new objects. They create new ontological entities that then become platforms on which we build new social networks. The artwork is an object but an odd one that has a foot in the subjective world of gist sensation as well as the objective world of external environment. In this, it is able to bring forth into being a new entity that has as its constituent parts the initial chosen (subjective) seemings, which combine with the framework of the (objective) zeitgeist and the (objective) conceptual definitions that would apply to any represented and linguistic objects. It now concatenates a new object. The earlier example of Rachel Whiteread articulating the space under the bed and making that space into a new ontological entity is just one example, and it is a relatively easy example because it is a geographical space that is being pointed to. But all art does the same thing.

Art is a way of navigating the plethora of data that comes to us—that sandstorm of ground-level ontological particularities as well as the larger semantic ones—and gives the artist a saliency map that is distinctly individual. This individual map is in addition to the saliency map that all of our visual apparatuses have that are grounded on such basic neural mechanisms as contrast, illumination, line orientation, etc. (This universal kind of saliency map has been discussed in several different places in the book.) But the saliency map that an artist develops—often called the artist's "practice"—is a map that garners information distinctly representative of that individual artist's viewpoint. This is where the idea of an artist's individual "style" emerges. The habitual choices become in-place filtering systems which then become automatic saliency maps. Thus, art gives artists a way of re-making their own world in accordance with their chosen seemings, which are dovetailed with the framing provided by the zeitgeist. And those chosen seemings (along with the zeitgeist framing) is offered as a new entity to the viewer. The artwork joins the constituents of the world as a new entity, thus remaking the world.

This is not terribly dissimilar to science. For science, by stipulating new kinds of basic entities such as subatomic particles governed by probabilistic existence, collapsing packets of probability waves, etc. gives to us a new framing of reality: Our description has changed, and so have the constituent parts of that reality. Art does the same.

Let me say a little about the ontological character of the constituents of those seemings. I have identified these ground-floor entities and have referred to them as p-unitates, and related them variously to the constituents of such notions as the low-level features in visual science as well as to the nonconceptual seemings in Evans. Let me be more precise in describing them. First of all, in my view, we want the primitive of our ontology to be a determinate one, and a phenomenal one, instead of the sort of indeterminate one that was offered by Quine in his value of a variable.

Sellars made this point, and I think it is a good one.[10] But though we want a phenomenal one, we do not want a phenomenal one such as what was proposed by Goodman, because that notion of a quale as a particular color, time, and place, was consequent to his metaphysics that forbid classes and led inexorably to his aesthetics which gave a referential account of a semantic (after the fact) analysis of artmaking.[11] In regard to the latter, instead of giving a semantic and referential account of the painting *after* one has viewed and absorbed the artwork (which is just a linguistic description of the artwork), it makes more sense to try to understand aesthetics in terms of what happens *at the moment* both when the viewer has the perceptual/phenomenalistic experience of standing in front of the artwork *and* when the artist is creating the artwork from the perceptual/phenomenalistic experience that was the original source of the aesthetic content. In regard to the former problem, e.g., that Goodman's qualia followed from his metaphysics which forbade classes, we could join Goodman is refusing to see particulars as instances of universals, but if we instead countenance classes, then we can, as others have advocated, see particulars as instances of a class (Sellars, 1952). This allows us to not only to take a more extensionalist approach to definitions (allowing for a usage that determines meaning option), it also leaves aside platonism, and it is consonant with vision science that tracks the members of saliency maps and the role of low-level features as instances of a class and not as instantiations of a different metaphysical order. It also has the added benefit that this members-of-a-set model is the inductive language learning models e.g., LLMs. Not insignificant pluses.

Therefore, p-unitates are a kind of knowledge by acquaintance, given to us in determinate moments, and processed through saliency maps. Artists' aesthetic experience is the paradigm of this perceptual cognition, because the artist-driven view gives us the elemental particularity at a level that is in the realm of flexibility of belief, and thus in the realm of creativity. Looking at the way artists see is a way to conceive of perceptual cognition that allows for the existence and role of the nonconceptual seemings and low-level features of perceptual experience and thus to see the world in a way that can be something other than the way it "seemed" initially. It is to open up not only the possibility for creativity but for social change.

But before we draw the curtain on the current chapter, let me say a final thing about two things. First, a final plug for Gareth Evans. Although Evans did not have the advantage of visual research into gist (Mary Potter, the first researcher in that field, began her research in the mid 1970s, and Evans died in 1980), his theory confirms that research, and even more perfectly dovetails the Chariker, Shapley, and Young research that argues for the heavy lifting being done in the visual cortex where the matching functions occur. Meaning happens here. It is the past experienced by that individual that creates these contextually-driven associations, affordances, and hence meanings. Secondly, I have only gestured toward the ontological

[10] See Willem DeVries' entry on Sellars in Stanford Encyclopedia for an extended discussion of this. (deVries, Willem, "Wilfrid Sellars", *The Stanford Encyclopedia of Philosophy* (Fall 2021 Edition), Edward N. Zalta (ed.), URL = https://plato.stanford.edu/archives/fall2021/entries/sellars/)

[11] See my Nominalism and its Aftermath: The philosophy of Nelson Goodman (Shottenkirk, 2009).

character of the artworks because, first of all, the debates regarding that are complex and extend beyond the confines of this book. This book is focused on art in terms of the functional aspects of (1) how we epistemically perceive the world as viewers, (2) how artists epistemically view the world, and (3) how the artworks made by those artists then function as conversations with the viewers, encouraging viewers to parse their reality anew in the same way that artists are taught to see reality anew. I will only repeat what was said earlier that artworks have a foot in both subjective as well as objective experience, which makes them unique as physical objects. They are a method of transduction from one to the other. What their detailed ontological character is a question for future researchers.

References

Albers, J. (2013). *Interaction of color.* Yale University Press.

Comrie, A. E., Frank, L. M., & Kay, K. (2022). Imagination as a fundamental function of the hippocampus. *Philosophical Transactions of the Royal Society B, 377*(1866), 20210336.

Cussins, A. (1999). Subjectivity, objectivity and frames of reference in Evans's theory of thought. *Electronic Journal for Analytic Philosophy.* http://www.phil.indiana.edu/ejap/ejap.html.

Evans, G. (1982). *The varieties of reference*. Oxford University Press.

Freeman, J., & Simoncelli, E. P. (2011). Metamers of the ventral stream. *Nature Neuroscience, 14*(9), 1195–1201.

Green, H. (1993). *Loving; Living; Party Going*. Penguin.

Greene, M. R., & Oliva, A. (2009). The briefest of glances: The time course of natural scene understanding. *Psychological Science, 20*(4), 464–472.

Groen, I. I., Ghebreab, S., Prins, H., Lamme, V. A., & Scholte, H. S. (2013). From image statistics to scene gist: evoked neural activity reveals transition from low-level natural image structure to scene category. *Journal of Neuroscience, 33*(48), 18814–18824.

Itti, L., Koch, C., & Niebur, E. (1998). A model of saliency-based visual attention for rapid scene analysis. *IEEE Transactions on Pattern Analysis and Machine Intelligence, 20*(11), 1254–1259.

Itti, L., & Koch, C. (2001). Computational modelling of visual attention. *Nature Reviews Neuroscience, 2*(3), 194–203.

Kneller, J. (2007). *Kant and the Power of Imagination.* Cambridge University Press.

Lettvin, J. Y. (1976). On seeing sidelong. *The Sciences, 16*(4), 10–20.

Lewis, R. G., Florio, E., Punzo, D., & Borrelli, E. (2021). The Brain's reward system in health and disease. *Circadian Clock in Brain Health and Disease*, 57–69.

Livingstone, M. S. (2000). Is it warm? Is it real? Or just low spatial frequency? *Science, 290*(5495), 1299.

Pierce, C. S. (1877). The Fixation of Belief, *12*(1), 1–15. https://philarchive.org/rec/PEITFO.

Schmidt, E. (2015). Does perceptual content have to be objective? A defense of nonconceptualism. *Journal for General Philosophy of Science, 46*, 201–214.

Sellars, W. (1952). Particulars. *Philosophy and Phenomenological Research, 13*(2), 184–199.

Shottenkirk, D. (2009). *Nominalism and its aftermath: The philosophy of Nelson Goodman.* Springer.

Stalnaker, R. (1998). What might nonconceptual content be? *Philosophical Issues 9*(1998), 339–352.

Tatler, B. W., Baddeley, R. J., & Gilchrist, I. D. (2005). Visual correlates of fixation selection: Effects of scale and time. *Vision Research, 45*(5), 643–659.

Torralba, A., Oliva, A., Castelhano, M. S., & Henderson, J. M. (2006). Contextual guidance of eye movements and attention in real-world scenes: the role of global features in object search. *Psychological Review, 113*(4), 766.

Chapter 5
Conversing: Art as Conversation

Abstract This chapter argues that the function of art is communicative: Humane conversation. Making art and perceiving it is not about object recognition, representation, or propositions. Instead, because it is reliant on gist perception, peripheral vision, nonconscious perception, affordances, and probabilistic cognition, it is a fluid response to the beliefs of others. Because the world comes to each of us solipsistically, our understanding of the world is necessarily a parsed understanding; we are *very* partial understanders. While the basic mechanism of our epistemological experience is the ability to edit, sort, and categorize the uncountable number of discrete bits of information, artists in particular are able to experience parts of reality in a direct, un-codified, and nonconceptualized state. Our bodies have evolved to know how to read low-level features instantaneously and are thus receptive to the non-linguistic attitudes and views of others in general. Thus, art is evolutionarily effective as a form of conversation. This is because art is an embodied form of perceptual conversation that results from consensus building and is constituent of the epistemological basis that forms public language—a way of conversing with one another that is embodied and not linguistic. Art is a way out of our solipsism.

Keywords Communication · Conversation · Humane · Nonconscious perception · Peripheral vision · Gist perception · Reggie burrows Hodges · Agnes Martin · Cecily Brown · Qualia

5.1 Introduction

This chapter combines the last four chapter, which were an analysis of art from the perceiver's point of view, as well as an analysis of art from the artist's point of view. We now can look at how art is a kind of epistemic conversation that allows us to empathetically understand one another and to thus ontologically re-make our shared social reality. This practice of epistemic conversations tell us how to chunk reality. Art begins with the artist's reliance on gist visual experience, which has nested

D. Shottenkirk, *Art as Cognition: How Gist Reframes the Aesthetic Experience as Conversation*, Synthese Library 521,
https://doi.org/10.1007/978-3-032-00143-6_5

within it sensorial and nonconceptual experience; this builds on the perceptual cognition of the perceivers and their similar reliance on gist with its cognitive processes that included peripheral processing, nonconceptual sensing, nonconscious vision, and the important psychological category of affordances.

5.2 Making Meets Perceiving

5.2.1 The Artist's First Person

When I begin a painting, I have already thought about it for some time and have a fairly clear idea of what I intend to do. Not all artists are like that. Some begin with the gut, so to speak, and just intuitively respond to the scene and/or the canvas, believing that thinking about it beforehand would inhibit those gut-level reactions and hence make it less authentic. Others, often those with a less subcortical intent in mind (I include myself in this), plot out the process in order to guarantee the outcome. But in either case, a common point remains: Artists place their first person account of the world onto that canvas. Let me explain what I mean by that.

The definition of a first person account is uncontentiously and has been previously referred to many times in the course of this book: The experience of reality from the point of view of "I", the "what it's like", instead of reality from the point of view of others. For example, the first person account of the pain of my toothache would be from my point of view, *my* pain, as opposed to that pain as described, objectively, by the dentist. The latter is the third person account.

For an artist, honing in on one's own first person account of the world is the fundamental and first task and is at the heart of all subsequent choices that are made. In other words, when a student-artist is first defining his or her practice, the first and fundamental question is "How do I experience the world?" In this, the student articulates both her sense of herself—her first person experience—but also *how the world seems to her* based on that first person experience. The two are flip sides of the same coin. Reality—or more precisely, one's definitional construct of "reality" or how the world seems—is consequent to one's articulation of one's own first person experience *of* that world. For example, if the student-artist experiences the world paranoidly, then the world itself is so defined and thus seen as frightening. Think of Edvard Munck's "The Scream", and his well-known battles with depression. That is a simplistic example, but in general that is the algorithm that is followed. Thus, the development of her style e.g., her "practice", is the act of putting that perspective onto the canvas. The same holds true for the mature artist, whose first person perspective is fully articulated and whose artwork is honed to accord with that. That is the essence of an artist's signature—what makes the artwork distinctively theirs alone.

This is done not just with subject-matter choices (e.g., do I paint landscapes or do I paint fantasy pictures of zombies?) but more importantly it is done with decisions relating to one's craft e.g., with the way materials are employed. For example, with my own work, it is easy to see how my choices in regard to materials reflect my subjective sense of the world—i.e., how I subjectively experience that world and thus how I take "the world" to be. This mirroring of my own first person experience begins at the most basic of craft decisions i.e., what kind of paints to adopt. I use oil paints because by adding Venice turpentine, Damar varnish, and linseed oil to the pigments, the paint colors become transparent and can then layer, creating hazy and translucent images. The communicative result of one transparent coat of paint being layered on top of another is that it creates a sense of contemplative mental thoughts instead of depicting specific physical things; thus, it can reference a different metaphysical domain (Fig. 5.1).[1]

To give another example, I eschew a thickly painted surface that covers the entire canvas as that surface carries with it the experience of muscular physicality and emotional bravado, along with less analytic and cognitive examination. Again, that would not be my choice as that is not my subjective reality, not my first person experience; I don't meet the world with those subjective responses that simplify, emotionally react, and take control. Instead, because I experience the world as always cognitively complex and often ambiguous, I react differently. My first person experience is to feel the complexity and opacity of reality (sometimes with bafflement, sometimes with amusement, sometimes with anxiety) and to step back and reflect, bringing order into the scene instead of responding with emotional bravado. But, at the same time, I also experience the visual world as richly seductive, with textures and colors conversing sometimes in sync and harmony and sometimes antagonistically. For example, a cream color can be on smooth, glassy silk and it can be on a bristly, coarse-haired boar. And it is the same cream! This visual conundrum that is part of daily life leads to the same first person experience: The world is unpredictable and my subjective experience of that is to feel its instability; the world is also often inconsistent, illogical, and sometimes un-mendable. Thus, worry is there too in my first person experience as is, sometimes, aggression. The same has been noted about my work.

It is therefore the case that my experience of the world gets transduced into my art through, largely, the choice of materials. As it does with all artists. This is why artists spend so much time talking about these things instead of talking about linguistic object referents. I have heard artists, when talking among themselves, much more often discuss choices of materials ("I really love conte crayons instead of charcoal pencils because the charcoal pencils are too soft and can't capture the detail, and it just ends of looking too romantic") than I have heard them discuss

[1] Traditionally, acrylics could not do that; they lay flat on the surface creating a less 3D texture, and through their plasticity speak a more superficial and skin-deep language (pop artists were smart to adopt this material). (Though there have been some new acrylic products though in the past few years that give more transparency.) It is hard to, so to speak, *fall into* them. Acrylics don't allow for that; they keep the viewer's eye and body at more of a remove. So, I don't use them as that is not a reflection of how I experience the world.

Fig. 5.1 Dena Shottenkirk. *Gist Perception,* 2021–2. Oil on canvas. 24 × 24 in. canvas. 12× 8× 6 in. wooden box and slips of paper. (Courtesy of the artist)

choices of subject-matter.[2] The formalist decisions regarding the type of paint, type of brushes, colors used, surface textures, etc., are all so crucial for the very simple reason that they convey the essential first person perspective: The "what it's like" of experience.

Much has been written about first person accounts, with Nagel's seminal 1974 piece "What is it like to be a bat?" (Nagel, 1980) the go-to point of reference. As a straight-forward though (oddly) gentle attack on material reductivism and

[2] Although artists, when talking to non-artists, will often discuss the subject-matter of their work e.g., "I paint figures engaged in erotic activity", more than they would discuss their craft. The reason should be obvious: It gives the non-artist something to reference, to point to, to imagine. It is one of the clear advantages of reference (and extensional definitions) over meaning. Of course, ease of use doesn't mean correctness or truth.

physicalism, Nagel's article reminded people that there was something non-reductive about one's own inner awarenesses of the world, and that those inner dings of subjective reactions to one's place in the environment were the very essence of consciousness. Whether or not those subjective, first person experiences might someday be explained in physicalist terms remained unknown, he noted, but to claim reduction as a fact within reach is to bank on a promissory note that promises default. This is still true today more than a half century later. We simply don't know how it might be that the neurological stimulations from the environment that we can track from the outer world into the complexities of our brain might translate into that experience we call "first person". How, for example, does the light, as it travels from the lakeside and converts to electrical impulses, then moves through my retina into the ganglion nerves and on into the LGN, ultimately finding its way to the visual cortex, *then* translate into my experience of delight and repose? Where did the delight and repose come from? And at what exact moment in that neurological process did *it* appear? And what did it look like? What are the mechanics, the atomic structure, etc.?

These are conundrums, and we don't have the answers. But we have something else instead: We still have the robust notion of first person experience, and that is what we need at this moment. For the acknowledgement of that experience gives us the ability to both value it and to see its functional role in larger epistemic projects. And the main epistemic project is art. The reason for this is as follows. First person experiences, which seem to be imprisoned within the confines of our own bodies are, though, not necessarily trapped within us, forcing our own solipsism and remove from others. Quite the contrary. For art itself depends on the ability to communicate first person experiences. And it depends on it because art is the artist's transducing of his or her own first person experience. Artists, in this process, are not primarily placing objects in the world on the canvas; rather, they place *themselves* upon the canvas.

5.2.2 *The Perceiver's Phenomenology*

To understand how a first person experience gets communicated to others in an art context, we need to examine the process of the perceiver experiencing a work of art. Then we can be led to understand how that experience is the very linchpin for so much that we value as individuals and as collectives, and how conversation enables that experience. Let me therefore give an example of another kind of painting and therefore another kind of encoded first person experience, and also—differently though this time—from the point of the perceiver instead of the maker. In the last sub-section, I used my own experience in the analysis of the first person perspective from the point of the view of the maker of art, but in this sub-section I will use my own experience as the perceiver of an artwork in an analysis of the *perceiver's experience of that artist's first-person perspective.* At the risk of appearing to over-simplify the choices, I will refer to a painter who encodes muscular physicality and emotional bravado: Cecily Brown.

5.2.2.1 Description: Looking at a Painting

In a gallery in the Tribeca section of New York, I stand in front of a painting; the fluid and rapid brush strokes in the vertical canvas flash at me blurred-style. Brown's *A conversation piece* is a relatively small canvas and yet, like all of her work, it explodes (Fig. 5.2).

The strokes are long and unhesitating, each soaked with generous amounts of paint and solvent, wetting the surface and making the paint slide like tires on an oil slick. The paint runs between control and abandon, but there is structure and composition here, too, as the pink swish at the top of the canvas is held in place by the golden yellow band at the very bottom. It is a frenetic world that is also cleverly ballasted by two vertical lines (trees, maybe or fence posts) at the top, and at the

Fig. 5.2 Cecily Brown. A conversation piece, 2023. Oil on linen. 29 × 17 in. (© Cecily Brown. Courtesy Paula Cooper Gallery, New York. Photo: Genevieve Hanson)

bottom of the canvas (exactly equidistant between the verticals at the top) are the starkly contrasting black outlines of pentagon interlocking shapes—pavement tiles, presumably. I see—only after a moment—two dogs stationed in the central part of the canvas, but their presence is almost obliterated by the wet-looking and smeared streaks of paint as those brushstrokes traverse the canvas, setting up a frantic speed as they travel the narrow distance from right to left, left to right. And though the greys dominate, they do so as many species of grey: some pink-grey, some cinnamon-grey, some yellow-grey, some brown-grey. All of these join forces creating a general tenor that is an odd combination of semi-violent movement with the hilarious banality of the stately dogs in the middle, seemingly oblivious to their whirlwind world.

As a painter who has reached back to expressionism, Brown epitomizes the confidence and unfurled passion that speaks with heated vibrancy and unleashed instincts. This is her practice: A practice that has been noted as breathing life into the tradition after the post-modern neo-expressionists of the 1980s were thought to be the last gasp. The point here though is not historical, but rather philosophical. What I have described is *her practice*: her point of view, her honed message, the *meaning* inherent in her work. It is thus not reference that is at issue here, but meaning. The artist's practice—those honed skills that are habitual and procedural memories—are the encapsulation of the artist's first person experience.

It is pertinent here to reiterate some points that we saw in the last chapter. Firstly, artists are not focused only or even primarily on linguistic objects. It is *the way* something is done that dominates decisions; the subject matter is only the starting line of the race. Secondly, and consequently, the primary emphasis is on formalist issues like negative space, peripheral parts of the canvas (and how that interacts with the area of central focus), color choices and surface textures, brush strokes, paint thickness, etc.

Obviously, these formalist concerns e.g., color, textures, paint choices, etc., are encoded in low-level features and are given in gist experience. Let us go back for a moment and reflect on the Brown experience and then, separately, to remember the basic process that the artist goes through. In the experience of seeing Brown's painting, my initial gist would have registered both the colors (greys, mostly, with some yellow and black and pink), the very rapid and blurry shapes, and rough textures. Not static, but fast moving though not disorderly. (If you squint your eyes and look at the reproduction, these low-level and low-resolution traits will be obvious.) That's an enormous amount of information and looking at the painting again in a foveating manner reconfirms that gist information, rather than contradicting it.

Now to revisit the basic process that the artist goes through in making art. The artist experiences the world through re-iterated gist perceptions, and notes (both consciously and nonconsciously) the first person experience embedded in those visual experiences—"This is how I feel about the world as it comes to me". But now it is more complex as we have, in addition, also switched over to the experience of the perceiver e.g., when the artwork leaves the artist's studio and goes out into the world at large and is seen by viewers i.e., by perceivers. So, in fact, now we have *two* gist experiences that are epistemically functioning as ways of gaining knowledge: The artist's gist and the art viewer's gist.

5.2.2.2 Analysis: Two Qualia Experiences Have Combined

For it is the artist's gist that, once re-packaged and given over in physical form on the canvas, is now fodder for the epistemic experience of the viewer, an epistemic event which is also structured around reiterated gist perceptual cognitions. The perceiver experiences the low-level features of the artwork as gist, and then—with each reiterated whole-scene perception that follows a blinking or re-looking at the artwork from a different perceptive—a new gist experience occurs. Over and over. We look and relook, with eye saccades governed by saliency maps, building up an accretion of impressions that then sum together to form an experience.

Let us think for a moment of what it meant to experience all the varieties of grey in the Brown painting. What meaning did they carry with them and how was it accomplished? As a side point it is useful here to remember the distinction between property attribution and the experience of a property in an aesthetic context. An aesthetic experience is locatable in percepts that do not necessarily succumb to a subject-predicate relation or a propositional that-statement. For example, when we look at red in the instance of trying to choose a grey shirt in a department store we experience the feature of grey differently than when we look at it in an aesthetic experience, say when we are looking at grey in a painting. If we think of ordinary tasks, such as in the example of picking a grey shirt, the grey is targeted, named, and linked to that object. The shirt is grey. There is a clear subject/predicate relationship: The grey belongs to the shirt. This is distinct from an instance where one is merely *aware of* the predicate grey.

This important difference can only be understood if we think about the role of gist in the aesthetic experience. When I experienced the multiple greys in Brown's painting, I didn't experience those greys as belonging to any object; they were not predicates of a subject, even of the subject "painting" or "canvas", and they were not proprietarily owned. They existed as distinct and independent entities—as low-level features.

They had been freed, so to speak, and were on their own. This is one of the central features of the aesthetic experience.

This feature then combines with the epistemic function of accessing the work in terms of the first person experience of the artist. This is another facet of aesthetic experience that distinguishes it from ordinary gist experiences. For in an aesthetic experience the first person experience of the artist has been given over to the viewer, who then has their own gist experience *of the artist's first person experience* and is therefore able, at least to some degree, to re-create and re-experience *the same first person experience that the artist had.* For example, when I looked at Cecily Brown's painting I felt invigorated, freshly and powerfully impassioned, reckless, and fearless. At that moment, I psychologically responded to the world with at least a thrust chin if not full-on chest-thumping. This process is a kind of channeling; I have channeled Brown's first person experience and in that process have come, at least to some degree, to experience the world as she experiences the world.[3] I experience her first person experience.

It is thus that makes art a kind of epistemic conversation that allows us to empathetically understand one another. Because the perceptual cognitions that constitute the aesthetic experience are a transduction of the artist's first-person experience now given over to the perceiver, meaning is derived from this experience.

The meaning is the "what it's like" perspective for the artist which has now been handed to the perceiver, with the perceiver now knowing a perspective of the world that is distinct from the perceiver's own. It is an additional perspective on the world. The artwork means something by virtue of encoding that perspective. Now, the perceiver's world has been enlarged to include the framework of someone else they have never met, and will probably never know. And yet knowledge is there.

5.2.2.3 Perception: The Basis of Thought

In this, that meaning is not completely reducible to linguistic description.[4] Words can begin to point to some of the phenomena, but cannot fully translate the first person experience or its constituents in the low-level reiterated gist. Thus, what can be captured by language is not the same as perceptual experience. I take it that this was also Evans' point when he was arguing that our perceptual experiences do not necessarily line up in one to one correspondence with our conceptualizations. His point is worthwhile summarizing for it is relevant to the issue at hand. In arguments against Kripke's causal theory and ultimately in defense of Russell, Evans was developing arguments for reference—and even more so for meaning—on the basis of perceptual experience, not on the basis of structured reference within an ideal language. And that perceptual experience is initially founded not on conceptual content but on bits of discrete perceptual data — that, while they might in the future be part of concepts, are not necessarily so in the present moment of the experience. In other words, he was opposed to Frege's ideal language requirements and wanted to instead figure out how it is that we in fact communicate, even in those instances where reference is blurry. For we are often vague about reference and yet we still have meaning. How is that accomplished? Evans begins with his criticism of Frege regarding this:

---

[3] This is certainly not to say that Brown goes through the world chest-thumping. I don't know her, but regardless of that, the first person account that is encapsulated in artwork is not the same as one's personality. How one responds to the world—how it feels from the Nagel-type "what it's like" perspective—is recorded internally and not necessarily outwardly. The artist's inner life, which is what is displayed in the artwork, is not usually what is displayed to the public, although notable exceptions of people like Ernest Hemmingway come to mind. More representative ones are Jackson Pollock who presented to others as mild-mannered or Max Beckmann, whose paintings were vibrant, sometimes violent, always highly activated and intense, but who, according to another spectacular artist from the same German era, Georg Grosz, had "the personality of a paperweight".

[4] Cf. Bermúdez, 2007.

> To summarize, then, I suggest that we take Frege's ascription of a sense to a Proper Name to mean that not only must one think of an object—the referent of the term—in order to understand a sentence containing it, but also anyone who is to understand the sentence must think of the referent in the same particular way. It is therefore, for Frege, as much a public and objective property of a term that it imposes this requirement, as that it has such and such an object as its referent (Evans, 1982, 17).

> Yet it appears that Frege was prepared to allow that there were perfectly meaningful sentences of the language, which could be used to express and convey thoughts, but to which his theory of Meaning would not apply. And he was apparently prepared to allow that parts of such sentences (specifically, empty Proper Names) could make a regular and systematic contribution to the thoughts expressed by sentences containing them, without having a semantic value of the kind he deemed appropriate for such expressions. This position seems barely intelligible (ibid., 1982, 23).

The criticism of Frege was that Frege both wanted reference to function universally in order to fulfill the demands of an idealist language (e.g., propositions) guaranteeing objectivity, and would also admit empty proper names e.g., Odysseus, which had no semantic value. Language though, for Evans, doesn't function like that. This was Russell's complaint against Frege, but Evans' response to Frege's problem is not entirely isomorphic with Russell, as Evans is also concerned about the not infrequent divorce, so to speak, between reference and meaning. In the following passages he points to the dilemma:

> Suppose, for example, that on a certain day in the past, a subject briefly observed two indistinguishable steel balls suspended from the same point and rotating about it. He now believes nothing about one ball which he does not believe about the other. This is certainly a situation in which the subject cannot discriminate one of the balls from all other things, since he cannot discriminate it from its fellow (ibid., 90).

> Does perception of an object always provide one with discriminating knowledge of it? (ibid., 91).

His point is that the structures of our language do not properly reflect our mental and—more importantly—our perceptual processes; we can be vague about reference, and yet we can still have meaning. In other words, the object is not under a definite description and yet the meaning is still communicated. The point is obvious: We can be vague such that the object can fail to exist in a third-person sort of way, and yet we are conveying meaning. His question then is how is it that we are standing in relation to that object? The answer was perceptually.[5]

This is a very difficult point to articulate, and though Evans didn't perhaps live long enough to fully accomplish it, he did point our attention in a legitimate direction: It is perception that is the foundation of thought, and that foundation is both independent of language and grammar and is not fully reflected in language or grammar. I take it that this is his essential point, and I completely agree. In this, perceptual experience is not completely mirrored in language; there is no

[5] See also Louise Antony's review of Evans in Philosophical Review (Antony, 1987): "*His goal was to argue a version of Russell such that some sort of 'direct reference,' non-descriptive and perceptually based, lies at the base of thought.*"

one-to-one correspondence. This is why Evans spoke of different "abilities" and also of nonconceptual information.

In closing, I would add to this that the nonconceptual component of visual experience functions in the construction of thought analogously to both qualia in perceptual experience and low-level features in gist. Each of these kind of perceptual experiences serve not only as referent-guides in the cognitive apparatus (i.e., they each function within the cognitive system through their ability to ostensively point to relevant features), but serve to also frame the terms of engagement through the function of sensing (i.e., they convey the meaning and value of the thing perceived). In this, they each give both the *what* that is perceived as well as the *how* it is perceived. (More is said regarding the relationship of (1) the meaning of the perception to (2) its referent in Sect. 5.2.3 in the discussion of J.J. Gibson's affordances.) They do this because they are fundamentally and ultimately first person experiences and thus have contained within them the emotional and psychological charges that, when transferred into either the conscious perception of objects or into nonconscious cognition, carries with it the valence, so to speak, of its original charge. Thus, it frames our discourse, it alters our perspective.

5.2.3 *Joint Affordances: Trans-qualia*

This, then, is art: The transference of the artist's non-representational qualia to the perceiver in the form of the artwork, which then becomes the fodder for the perceiver's qualia and thus aesthetic experience.[6] It is through this ability to carry a distinct valence that makes art a kind of epistemic conversation that allows us to empathetically understand one another. How is the understanding and the empathy accomplished?

5.2.3.1 Seeing through

Let me first briefly recap the mechanics of how it is that we understand the artist's communication before we move on, in the next chapter, to the related topic of empathy and its associated mental states. The "feel" of the painting is found in its formalist, low-level properties that communicate as qualia and carry with them emotional valences, which are not the sort of things that would necessarily be noticed in the day-to-day negotiation of objects as we move around things or proprietarily hand

[6] While qualia are generally thought to be non-representational first person experience and thus are not a representation of anything, it is the case that the qualia are represented *by* the artwork. To represent something is to stand for it; and while qualia do not stand for anything but are instead the intrinsic phenomena we have when we visual experience, it is therefore easy to agree that qualia do not represent and also easy to understand how an artwork might, on the other hand, represent qualia.

things back and forth between each other. *That* world is the world of named objects. Instead, art shows us more fundamental non-object things that give a point of view—analogously to a gist of the scene. The particular style or signature seen in an artist's work encapsulates a point of view that has now been handed over to the (observant) viewer. "I get where the artist is coming from" people sometimes say after they've seen an artwork, or "I know what that feels like" or "I feel like I've been seen". This is accomplished, as I have argued, largely through the formalist properties (low-level features) that both encapsulate and convey the qualia of the artist's first person experience.

And that is enormously powerful. I was once in Amsterdam and standing in front of Rembrandt's *The Night Watch (1642)* when I noticed that the woman standing next to me, a stranger, had tears running down her face. I looked at her and we shared a sympathetic facial exchange; words were of no currency. What is it about Rembrandt that has so moved people for the last almost 400 years? It is not the mere subject-matter—in this instance, military men gearing up, which very rarely brings people to tears. It is so hard to put into words, and I am befuddled by the difficulty for I feel so potently the quality of Rembrandt's luminosity as it is surely he alone (well, maybe Monet is here in this category, too) who was able to capture the ephemeral yet tactile quality of seen-light once it's transferred to pigment; one can almost feel the warmth of the illumination when it is bright and flooding a face as well as the coldness of the temperature when it has been eaten up by the shadow. But there is so much more to his work than that, and my description does not begin to capture the experience; surely, the wise thing is not to try to put it into words. For that is not its venue.

Rather, we can say with more confidence the philosophical (and worded) point that art is a way of giving a different kind of reality than what words give: A re-shaped reality. And the artist gets there by importantly *not* seeing something only in terms of the names that have been attached to it. It is seeing it apart from those names. The names will only get in the way of the experience.

Sensing is not merely having a named object, e.g., a noun, and an attached predicate. Sensing, in that way that one fully experiences the moment, is not just recognizing an antecedently defined object; it is also experiencing, for want of a better term, the qualia of the moment—the light as it hits one of the faces in Rembrandt's painting. The crisp texture of a white area, etc. It is much like the act of meditation.

It is seeing *through* the named objects and into the experience itself. And it is that that has the potential to redefine that experience for others.

What I have been describing here is how artists capture the gist of an experience and convey it others; in other words, how they capture the qualia. First of all, the artist translates his or her qualia into a new object. The artwork then exists as a physical object instantiating the first person experience of the artist, which allows us—the perceivers—however briefly or incompletely, to see the world as that artist saw the world. To revisit the Rachel Whiteread piece, I now see and value that space under the bed, and see in that normally un-named, private, and neutral geography a new realm imprinted by absented and vulnerable humans, a realm of things once there but no longer there, of time itself as it relentlessly erodes bodies. The sadness and embarrassment of an unmade bed multiplied by a factor of 10. Or to revisit Rembrandt's *The night watch*, the fractured pattern of blacks and whites across the canvas invigorates the viewing itself, which when combined with the many diagonals (of the banners, swords, guns, etc.) activates a gut-grabbing sense of excitement verging over to danger. The darks are too dark, and much is lost in them, and the precision of the details is too cold. I am left with contradictory emotions as I both am afraid of the organized battalion and excited by the movement itself.

Thus, the first-person qualia of the artist that constitutes his or her experience is transduced into the art object that represent that experience, which in turn is presented to the public, experienced by the audience and transformed, through conversation, into a social fact that conveys meaning. People talk to each other about what they experience, and that, in turn, develop meaning.

And central to the process is how it is that external objects and percepts come to be part of perceptual experience. The specific example served by the artwork is an analysis of how it is that we in general edit and construct external objects on the basis of first person experiences and how that first person experience can be adopted by others: The individual's point of view—experienced as qualia and translated as such—can and sometimes does become accepted by others and then becomes incorporated into others' third person definitions of a (newly created) object. In art, these are the new qualities of the object/art object that the artist has identified, and as I've shown in the case of Whiteread's sculpture, the qualities of experience—the qualia—can be constituted anew and hence a new object can be established.

5.2.3.2 Trans-qualia Not Concepts

Let me now say something about the paradigm that I am eager to dethrone. In arguing against a hierarchically dominating and governing role of concepts and instead in favor of the role of the nonconceptual founded in qualia; I am arguing for perception as the basis of knowledge. General terms and concepts can be easily over-rated as a pathway to truth when they are more properly functioning as a way speed is gained in the process of categorization. Looked at this way, concepts are transient terms, governing morphing sets of data, and are only functionally important as short cuts, much as short cuts on a keyboard are pragmatically useful. They are also ways to delimit and to proprietarily own, as we pass things back and forth between us much more easily when we have at our disposal general terms. We can say, for example, "give me the apple" and demand the apple, or say, "all dogs are mammals" and then sort dogs quite quickly. There's both convenience and ownership.

Instead of reifying concepts, we should instead remember that it is perceptual experiences that are the substructure of all experience. And they are different in structure from language. For perceptual data exist in a temporal order, and its use in thought is provisional. The temporal order not only bombards us with a plethora of data, but as has been noted, it does so unevenly as some bits of perceptual data are consciously recognized by us, others goes recognized only nonconsciously, and other bits of data are completely unrecognized. This triage is not only the consequence of socially named entities but also, as well, of our individual biases that trail as a consequence of what we allow ourselves to see and recognize, and what we filter out.

Concepts only attach to the public objects and are a reification of data that removes data from the informational, perceptual, and temporal experience in which it exists for the perceiver. Thus, concepts are an abstraction of information, an idealization of the process. As rigid constituents of content, concepts are only so by dint of their idealized nature. I think that was what was at the heart of Evans' objection regarding the notion that content presents concepts not only as the governing categories of experience but as being reified, monolithically culled, and purified of the messiness of ordinary experience.

But it is the latter—the messiness of ordinary perceptual experience—that provides us with the groundswell of our information. It is a dynamic system that exists in time. Art communicates within this system because of the nexus of qualia that exists in the social space between the artist's qualia experience and the perceiver's qualia experience. This *trans*-qualia is the space of knowledge acquisition. It is neither subjective—as in the subject—nor objective, as in the object. It is both.

I use that phrase with deliberation: Trans-qualia. To recapitulate the point: It is qualia that has been passed from the artist to the viewer. Now shared. Now jointly experienced, with affordances jointly shared.

5.2.3.3 Affordances

Let me now say something about J.J. Gibson's notion of affordances in more detail as it applies in this context. (Affordances were initially introduced in Sect. 1.3.1.) The notion of affordances argues that perception is direct, immediate and that that directness and immediacy is a conveyance of value, not objects. When I see something, what I am seeing is the thing's value to me; I want to either go toward it or go away from it. *Desire/aversion.* This is my body's response to the things in my environment. Thus, affordances are relational; this point cannot be over-emphasized. The usual definition is that it is what the environment offers the animal/human and thus describes the relationship *between* animals and their environment. For example, Gibson describes the relationship between surfaces and affordances in his seminal paper "The Theory of Affordances":

> How do we go from surfaces to affordances? And if there is information in light for the perception of surfaces, is there information for the perception of what they afford? Perhaps the composition and layout of surfaces constitute what they afford. If so, to perceive them is to perceive what they afford. This is a radical hypothesis, for it implies that the "values" and "meanings" of things in the environment can be directly perceived. Moreover, it would explain the sense in which values and meanings are external to the perceiver…The affordances of the environment are what it offers the animal, what it provides or furnishes, either for good or ill (Gibson, 1977, 1).

It is somewhat difficult to find the language to refer to the world in the way that Gibson describes. One can think of the Heidegger example of the hammer in his "readiness-to-hand" analysis, but that would somewhat miss the point. For Gibson, the point is not merely in how one can maneuver or interact with things of the world, but it is a fundamental re-thinking of perception itself that he is proposing. He states it thus:

> Orthodox psychology asserts that we perceive these objects insofar as we discriminate their properties or qualities. Psychologists carry out elegant experiments in the laboratory to find out how and how well these qualities are discriminated. The psychologists assume that objects are composed of their qualities. But I now suggest that what we perceive when we look at objects are their affordances, not their qualities. We can discriminate the dimensions of difference if required to do so in an experiment, but what the object affords us is what we normally pay attention to (Gibson, 1977, 7).

Gibson, I take it, is not only adopting the gestalts' point of view (which he addresses) but also the point of view of Hobbes: Wanting and fearing are our main motivating forces. Whereas Hobbes emphasized the fearing, Gibson emphasized the wanting, but the two are opposite sides of the same coin. To again quote Gibson:

> Kurt Lewin coined the term Aufforderungscharakter, which has been translated as invitation character (by J. F. Brown in 1929) and as valence (by D. K. Adams in 1931; cf. Marrow, 1969, p. 56, for the history of these translations). The latter term came into general use. Valences for Lewin had corresponding vectors, which could be represented as arrows pushing the observer toward or away from the object (ibid., 1977, 11).

We go toward something or away from it depending on our assessment of its value to us. And these are predictive and regular assessments; I will always go toward food and away from the wolf e.g., away from being food. The world itself, thus, is structured and presents invariant phenomena that allows species to adapt.[7] This view, which relies on invariance in the environment, incorporates then a kind of universalism, but it would be the sort of non-platonic universalism advocated by William of Ockham, who wrote that crying was a universal sign of weeping; but that was the only sort of universals he would countenance as it didn't fall afoul of his nominalist commitments. For Gibson, too, such universals are simply in the structure of biological laws and we are, fortunately, able to perceive them, and thus we can maneuver and adapt.

How then are the low-level P-unitates that we perceive in artworks—e.g., that blob of grey there or that hard-edged texture here—instances of affordances? It is very simple: They are those direct, immediate non-objects which either repel or entice us. They communicate through their value to us; they communicate to us because their values are perceived immediately and directly. Gibson entitled one

[7] See Shaw and McIntyre on the requirements for a structured environment: "What are the general conditions that make knowledge of physical events possible? First, if the world consisted of an unstructured chaos then no knowledge would be possible. Even if the world consisted of a plurality of uncorrelated events, the exigencies of life (assuming it were possible) would be totally arbitrary, thus precluding adaptation by any living creature. For knowledge to be possible, then, we must assume uniformity of structure at some level of analysis across all classes of events, and assume that such structure is preserved to some significant extent by the knowledge-gathering processes of organisms" (Shaw & McIntyre, 1974, 326).

subheading, "To Perceive An Affordance Is Not To Classify An Object" (Gibson, 1977, 7). This couldn't be said more clearly. When we look at an artwork, our emotional valences are triggered by the features that I have alternately labeled low-level (a term borrowed from vision science), formalist features (a term borrowed from artists and art historians), qualia (the philosophical term for the entities constituting first-person experience), nonconceptual content (the philosophical term used in contrast to the content in conceptual perception), or P-unitates (my own preferred term, which incorporates the former). While each of these terms not only originate in different disciplines and varyingly refer to slightly different sets of entities, their overlap is clear, and they are all instances of non-objects that would elicit affordances.

I said earlier that the trans-qualia is neither subjective—as in the subject—nor objective, as in the object. This is similar to Gibson's notion of the affordances, which he states "Affordances are properties taken with reference to the observer. They are neither physical nor phenomenal" (Gibson, 1977, 17).

This in-between world is a social world and it is not unrelated to the distinction between conceptualized and nonconceptualized.[8] The latter plays an important role in how art is created by individuals and understood by viewers. While concepts enter into a description of how the viewer experiences art *after* the moment of experiencing the art, they enter in less so in both (1) the actual moment of the viewer is experiencing the artwork as well as (2) the artist's capturing of his or her first-person experience while making the artwork. It is part of the contention of this book that artists in particular are able to experience parts of reality in a direct, un-codified, and nonconceptualized state. Seeing things differently, establishing relations where others had not previously seen such relations, being aware of the particularity of each bit of reality in its un-codified and nonconceptualized nature is what it means to be creative. And it is an epistemological act that is necessary if there is ever to be creation or change.

5.3 Perceiving "The Other"

5.3.1 The Body

As art is the artist's identity worked into a practice, worked into an object that is then a re-creation of what "the world seems like to me", it is a re-creation of the qualia that can now be handed over to someone else, so that they too might experience that qualia pattern. Art is thus the way that a person—e.g., the artist—isolates their own subjective experience of the world, which as subjective experience can be defined as instances of qualia, and then translates that experience into a physical

[8] There have been many ways of formulating this distinction; for example, early Indian philosophy's distinction between unconceptualized (nirvikalpaka) or conceptualized (savikalpaka) perception.

object that re-creates the same, or at least similar, subjective experience in the viewer. Examining exactly how this is done is a window onto the mechanics of social object construction.

This works, in part, because of the geographical peculiarity of experience. The artist's experience begins with the perspectival solipsism that joins the "I" and the "here"—they are causally connected. I get to be an "I" because of my awareness of "here".[9] This should be a familiar story. I/my body stand here and look out onto the world; this was also a point emphasized by David Marr. I absorb the world through the porousness of the body that is located in here/now; the world enters through those sensing mechanisms that we call our eyes, ears, etc., in a way that makes the world funneled—it narrows down as it meets those specific porous points of entry and thus makes the body absolutely centric to the experiencing of the world itself. The world comes to me, and it comes to me in a funneled, pin-pointed way that forces my importance onto myself; my experience is singular and without comparison, and in the silence of others' experience, my experience becomes univocal. *This is the way the world is.*

The first person perspective is thus an awareness of (1) my own body as the centric point to which all incoming data is directed and by which all data is measured—the "here", and (2) my awareness of the "my-ness" of my senses' mechanical parsing of the qualities of that data we call experience (e.g., to be aware of seeing, of hearing, etc.)—an important part of what constitutes the "I".

Regarding the latter, it is *that* that gives first person accounts their proprietary nature. It is *my* experience of the red leather ball as it bounces across the room, *my* awareness of the redness as a distinct quality, *my* awareness of the thudding sound of the leather as it hits against the wooden floor, *my* awareness—whether it was fun or threatening, chaotic or boring. My awareness of my love for a loved one, my experience of an artwork, or my own pain are all the very essence of what are proprietary to me, despite being awarenesses that are difficult to translate into language. Likewise it is *my* awareness of the complexity and contradiction of the world that makes it my first person experience, and it is I who determines whether that is fun or threatening, chaotic or boring. The "I" that is drawn from the my-ness of my sensory experience is a result of the my-ness that is drawn from locating my body in the here. Thus, first person awareness is geographically/temporally/spatially located, and it is thus proprietary. And when I stand in front of Cecily Brown's painting, I too experience a version of her qualia—her first person perspective. This is where the issue of joint affordances enters in. My gist experience of her artwork gave me the immediate sense of swerving and swooshing movement, excitement, verve. Values are perceived immediately and directly. This is art.

In this book I am arguing for a different view of both subjects and objects than is the traditional, separated view. In the view I am putting forth, our bodies, porous to the world, also construct that world. Not absolutely and without limits, but it is our editing and building that constructs what it is that we then deem as real. It is our

[9] See Evans on the general topic of "here" (Evans, 1991, 152–157).

bodies' sensorial absorption of information and the mental constructs that are consequent to that absorption that construct reality; it is a reality experienced piecemeal and its construction the result of social induction and consensus.

Thus, I am arguing against the view that advocates for the strict bifurcation between subject and object. Instead, I am arguing for the different paradigm of the body, one that is more porous to the external world in such a way that the divide between the body and the external world is not complete, resulting in a description of the subject as constantly being reconstituted by the absorption of the external data; and of the object as constructed by the phenomenological experiences of the subject. And thus our notion of "the real" shifts like sand, making our notion of "the objective" world merely the sum total of the vast majority of our individual views.[10] Consensus looms large in reality-construction, making reality fungible and constructed by us.

5.3.2 *Empathy*

Art is a dynamic, temporal system that is the exercise of recognizing invariant, low-level features in the environment and training our perceptual systems to respond appropriately the those stimuli i.e., learning how to recognize affordances and hence increase our valuing of them. I propose, in addition, that learning to seek these low-level features that draw one in on the basis of affordances also serves as training to recognize those affordances in other settings. For example, if one spent time looking at Degas' "Violinist and Young Woman" the eye will probably first go to the violinist as he is wearing a red jacket and all other colors in the painting are neutral grays, tans or browns.[11]

And yet, in an absolutely involuntary way, the eye darts to the woman's hair where an ornament of high contrast black and white sits nestled into her blond hair. The detail of the hair ornament in the Degas painting and its contrast with surrounding colors makes it even more of a central feature than the violinist. It is black and white, but extreme black and extreme white. Very small, but very powerful. I could almost feel those contrast neurons in my visual cortex start to shudder and shake! In addition, her dress, a brown-grey field, has a brown umber shadow that is painted in a slightly diagonal way (diagonals create a sense of slight instability in us) that runs all the way from the back of the dress on the left side of the canvas to across her lap

[10] This is true across disciplines—dark matter is now thought to be a fact of the world thanks to Fred Zwicky in the 1930s and Vera Rubin who, in the 1950s argued for it only to be ridiculed. This is of course the explanation behind the endless revision of science, for what is scientific "fact" changes according to the consensus view and that view is always open to revision. But it is true in our cultural world as well: Human rights exist in a way they didn't a hundred or a thousand years ago and are thus thought to be real and not merely imagined, etc. Reality is constructed.

[11] To view this painting in color and high resolution, see: https://www.metmuseum.org/art/collection/search/844685

and underneath her hand that is holding a sheet music book; it is more than loose brush work—it I more like a field of color but without gentle gradations or purposeful brushstrokes; in fact, it looks like he was smearing his brush in preparation for cleaning it! All these moments in the painting draw one in and in the process of looking one loses oneself in the act and continues responding to his instances of trailing brushwork. It dances, freely, without hesitation.

How was the eye controlled? As we saw in earlier sections of the book, some eye saccade movements will be controlled by saliency maps where such mechanisms as light/dark contrast control many of the choices, as well as colors that are placed next to contrasting colors. This fact is across the species and is a bottom-up mechanism. Gist, too, operates across the species with the low-resolution shapes conveying their spatial arrangement and establishing dominance relations; the red on the violinist was known at that moment, too, as was the hair ornament. Contrast and line orientation are central to this process. Furthermore, people often look at faces, as facial recognition is such a large part of our brain. Additional mechanisms are also in place for each individual though they are the result not of biology but of interests habitually practiced; for example, I am apt to look at the folds in the cloth as I have worked on such painting techniques myself.

How do these moments we know from vision science including whole scene gist and eye saccades factor into the notion of affordances? It is through the immediacy of their attraction and the value that they convey. When I saw this particular painting at the Met, I was with my daughter and I asked her which part of the painting did she think my eye first went to. She answered, "The ornament in the woman's hair". She was right: I had absolutely loved that little irrelevant detail in her hair that so aggressively claimed its own status. When I asked her why, she said it was because it was of high contrast, but also because it was so ridiculous that the part of the painting most likely to draw one's attention was also the least related to the narrative. This was true; the white in the middle of the ornament was like a light beacon, your eye couldn't help go to it and for me, the ridiculousness of that as the central feature was thrilling. I loved the defiant insistence of its importance and the war it had created with a narrative that could have been bland e.g., man playing a violin to a woman. Instead, the violinist had slumped into low-contrast obscurity, and she, with her facial expression of combined interrupted concentration and now annoyance as she has turned to look at the intruder who is now recording her life, does not convey a coquettish girl which one might expect in that situation in the time-period. And it was the little ornament that helped accomplish that. In addition, the ornament created an almost perfect equilateral triangle between it and the scratched shadow underneath her hand and the violinist's hand where he holds the neck of the violin, all of which are high contrast areas. Compositionally, the ornament is anchoring. It was not a mistake. Which makes it even more powerful. It brings seriousness to an image that might very well have been less without it. The affordance it provided was a very strong pull toward it, even though I could not decipher what exactly it was—a bow, a clip? It didn't matter. It did its job.

Thus, these p-unitates that we experience in art—e.g., the bright white next to the dark black—is seen with its affordances because it has its affordances. I was drawn to it, and knew somehow or other that it was useful to me. It satisfied needs of mine. And after an exploration of how it satisfied me, I developed many other more complex thoughts. Thus, art is the way physical objects can be the delivery packet for thoughts.

That Degas painted it thus is the first moment of transportation; he put it out into the world. The second moment of transportation is when it hits the viewer. The viewer sees the artist. It is then a greeting, a way of granting recognition, a way of saying "I see you"; "I care." The viewer says this, either to themselves or to others, in so many words, in the second moment of transportation. *This* I care about. This moment of transduction is the felt moment, the moment of inner moment, the moment of affordances. It is a profound commitment. The third moment of transportation is when the viewer talks about the artwork to others. It is here where art becomes incorporated into thought. This is the very unique product of art and it is addressed in Sect. 5.4. But for now, I would like to return to the second moment of transportation and focus on what it is that is happening in that phenomenological experience when we stand in front of an artwork.

Empathy is one thing that happens. Much has been said about this word and its distinction from the notion of sympathy. Suffice it to say, the distinction rides on identification. I had said at the beginning of this section, "that learning to seek these low-level features that draw one in on the basis of affordances also serves as training to recognize those affordances in other settings". What might one have learned from looking at this Degas? I would argue that allowing oneself to be drawn initially to the hair ornament and being thrilled by the insouciance of it trained a particular observational ability and a particular emotion: The most telling parts of a scene (or a person) are often not in the blatantly advertised parts, but (in the phrasing used by poker players) in "the tell"—that little part that someone lets slip, or that little detail in the scene that could easily go unnoticed but is the linchpin for the explanation. Careful observation is often best calibrated not by scanning the scene pixel by pixel but by letting your gut reactions to the affordances draw your attention. What are you repelled by? What are drawn to? Paying attention to those things allows for the summation process to calculate properly. If I'm playing poker with someone or negotiating a deal, I pay attention not to the consciously delivered and received smile but to the little bodily gestures, tones of voice, etc., that cause me to move away or move toward. This sensitivity to my own body's response is the key to empathy, and it is also the key to understanding. Art trains this.

5.3.3 *Trust*

Because art maintains us in a world of particular concrete experience, it conveys to each of us—as viewers—that particular embodied experience of the world that the artist has given us and at the same time conveys the appropriate interpretation of that experience. In other words, it tells us not only what to look at but more importantly *what attitude to hold* to what we look at e.g., how to value it. Thus, we look at the subset of data that the artist has chosen for us—that subset that is selected out from the sandstorm of data confronting us at any given moment—and in that subset we are given data that is legible and meaningful. For example, I look at the Degas and accept the world as slightly unhinged and freer that I might have expected (e.g., unpredictable brushstrokes, surprise moments of color and value contrasts, etc.). Additionally and consequently, I have an attitude of respect for the woman sitting there next to the violinist—I take her seriously for she (read: her personality) has been focused on and I in turn focus on her; the high contrast ornament in her hair and the whacky brushstrokes in the shadow on her dress were enough to encode that attitude. To take a different example, I look at the Cecily Brown painting and accept the view of the world that is exploding with life and slightly dangerous.

Very importantly, these are not instances of abstract knowledge. What art gives is, rather, a version of lived and interpreted sensorial experience. And as a particular artistic view takes hold in the world, more and more people now adopt that experience as, at least partially, their own.

The process of exponentially larger numbers of people accepting the view presented by an artwork is an example of the process of consensus in society. It is the process of reality building, and it happens consequent to trust. The viewers trust the artist and adopt the point of view. It is a shared attitude, a shared way of standing to the world, a shared set of values. This is the central part of "culture", writ large. People care about the art that has been produced from their culture because it embodies the culture. It embodies the values.[12] This is the central way that makes art a kind of epistemic conversation that allows us to empathetically understand one another and to thus ontologically re-make our shared social reality. These epistemic conversations tell us how to chunk reality.

Through-out this book, I have used the term "transduction" to describe this. It is now perhaps possible to discuss the terminology as we have come to the moment of reflecting on such issues as empathy and trust. The giving-over to another happens in the art context not in the normal gift-giving way or the object-transference way, as the giving-over in an art context is done on a more micro scale. Let me explain that a bit more fully. Transduction is a term used in science to mean the transference of material from one state or location to another, often used in physiology to refer to the transference of genetic material from one microorganism to another by a viral agent. It happens in a world unseen by normal human eyes, but a world that has rippling causal consequences down many chains.

Likewise, this is what art does. With art, there is a selection of material (e.g., qualia, formal properties, etc., coupled with subject matter) that is placed in the delivery packet called "an artwork", and it is handed over to the viewer, enabling the viewer to see—at least partially—the world from the artist's perspective. The viewer sees the world through the scrim that is the artist's reality. (The viewer is now infected.) This is why I think of the process of artists taking a subset of data and

[12]A culture's art is probably also valued because it shows instances of excellence and hence the culture's accomplishments. But this is derivative of the original reason.

making it meaningful as being close to what happens in physiology and thus can perhaps best be described as "transduction".

As recipients of others' creations, we see the world through the artists' transductions—through their sieve, as it were, as they sort the data—and hence are able to differently parse reality and understand reality itself more fully than before. We are able to parse reality differently because we've been exposed to the very direct epistemological choices conveyed by someone's art. We have been exposed to someone else's choices. Art matters because it gives us the pleasure of knowing others, ourselves, and reality itself.

But this knowing happens because we are in a moment of increased vulnerability e.g., we are open to the experience of someone else's viewpoint. But not entirely. It is like playing house; for when we experience art we can be vulnerable to others' views, feelings, and needs in a way that is only partially real. I say this with some hesitation as experiencing art is often the most real thing someone does. What I mean more precisely is this: We experience fully and openly someone's else's qualia—their first person experience of the world and thus their emotional commitments—in ways that do not fully entail the levels of danger that are entailed when we do so in a real-life situation with another human being. In the latter instance, the moment of openness to the other can, down the temporal line, (potentially) lead to disaster or pain, whereas in the artwork it does not. I can experience van Gogh without having to get too terribly close to the consequences I might suffer if I were roommates or lovers with him. Art, in that real-life way, is at a remove. Nevertheless, the experience of vulnerability itself is enough to create a sense of trust. And that is crucial, for it is the bridge to belief. (This is discussed more fully in Sect. 5.3.4.)

As art is not primarily the conveyance of linguistic or abstract knowledge, but rather the conveyance of a version of lived and interpreted sensorial experience, it is worthwhile to continue reflect for a moment on the related field of science and particularly on the biological and psychological bases that makes this trust possible. As art is a conveyance of the artist's qualia and first person account of the world, that conveyance must also be in accordance with basic biological patterns. What I mean by that is that the uniformity of meaning conveyed in art (e.g., static compositions convey harmony and order, diagonals convey disorder/excitement, cool colors convey calm, warm colors the opposite, etc.) is based in biology. These visual facts transcend cultural and generational divides as well as the constraints attempted by relativism. While the philosophical position of relativism has had its moments of popularity, the path to complete relativism is littered with corpses as writers have realized not only the factual truth that there are many things, such as the ones just mentioned, that are non-relative, but there is also the logical constraint imposed by the fact that if everything is relative then there remains nothing for things to be relative to, and ascertaining anything becomes problematic. Thus, we agree to certain Ockham-type universals. To again refer to Gibson's theories on this:

> Gibson's concept of information is dependent on the genetic preattunement of the species, and the education of attention for the individual organism, in the detection of certain fundamental invariants of stimulation from the terrestrial environment (e.g., gravity, sky-above and ground-below, horizon, texture gradient of the ground plane, texture-flow field proper-

> ties of the optic array, etc.). Such informational invariants specify important properties of the organism's environment, to which it must adapt to survive and maintain its health and well-being; hence, the perceptual information can be said to specify "affordances"of great ecological significance... Gibson assumes that perceptual processing of information is a continuous interactive relation between the organism and its environment. The all-important invariant relationships exist not in the discrete cross sections of the information flow, but in the constant longitudinal relations (Shaw & McIntyre, 2024, 309).

Our bodies are attuned to these basic contingencies of our environment and much of the information transferred from the artist to the perceiver relies on this "preattunement of the species"—i.e., I know what diagonals mean (something is falling! look out!) as well as I recognized the relevance of texture fields and color. Our brain registers the value of these bits of data from our environment and the emotional valences of those inputs. The amygdala is a central region for this, and the combined support of neurobiological studies show amygdala's role in emotional memory, with psychological studies suggesting that there is enhanced memory for the gist of complex stimuli (Adolphs et al., 2005), which would include artworks as well as other things. While all visual experience involves gist, the amygdala might have a special role in the more complex stimuli. The researchers Adolphs et al. conclude, "the data support a model whereby the amygdala focusses processing resources on gist, possibly accounting for features of traumatic memories and eyewitness testimony in real life" (ibid., 512).

Our brains register the importance of the affordances in both their detail and in their whole-scene gist, and that is a large part of how art finds its way into our lives. We are radically influenced by the emotion-generating amygdala, as well as the editing function performed by emotions themselves. These make us notice or ignore, care or not are. This is why I have repeatedly stated that we are constantly choosing and editing out, noticing and failing to notice, valuing and refusing to value. Choice is crucial here, and part of our choices are made by the acts of agreeing to be influenced by others.

The influence and trust is easier to understand if one again reflects on the biology of gist and their associated affordances. Because our bodies naturally rely on these universals referred to by Gibson, it is thus easier for viewers to trust the artists who in turn use these universals because they necessarily rely on them. For example, when I look at an orderly composition, I automatically trust that it is orderly (my body told me so), and I trust, therefore, the artist who told me it was orderly (telling myself, "This artist made an orderly artwork".) If I am drawn to that I will follow up with a statement like, "I like this artist." Or, if I look at a red painting, I automatically trust my bodily response to the redness (e.g., bold, passionate, etc.) and I trust, therefore, the artist who gave me the red (telling myself, "This artist made a passionate artwork".) The messenger and the message sync up and trust is increased. If I am drawn to that I will follow up with a statement like, "I like this artist." The deal is clinched. All nicely packaged.

In this, art takes on a particular and important role. Because art combines both "the what" to look at with "the how" to look at it, this complex role of art is instrumental in creating not only trust between the artist and the viewer, but trust that the viewer has for the larger point of view, now viewed as an abstract object. This process is incremental though probably something close to exponential as influences are passed from one person to others e.g., comments such as, "did you see the great Cecily Brown show?" creates the fertile ground for others' reception to that work. As Bruno Latour wrote, "We need, in other words, to look at the way in which someone convinces someone else to take up a statement, to pass it along, to make it more of a fact, and to recognize the first author's ownership and originality."[13] One person convinces a few other people of the value of some artist's view and if the convincing is spread far enough to enough people, the point of view is socially adopted. Thus, how art makes epistemological changes in the larger social consensus is seen through the incremental adoption by individuals of the legitimacy and truth of the particular view being presented by a particular artwork or body of artworks.

I would like to say one final thing in this subsection that is perhaps contentious. Many aesthetic theories ignore the role of the . I would argue that not only is that a mistake but that contemporary art in particular cannot function without an assessment of the artist's intention. And the reason for that isn't only that parsing the message of the art necessarily involves parsing what it was that the artist chose to hand over in the form of edited data (the choice itself necessarily an intention whether conscious or nonconscious). It is also because of the role of honesty. The latter has an important role in art as the artist's intention is only conveyed when it is clearly communicated. The latter can fail to happen, as is sometimes the case, if the artist has not honed or visually articulated his or her concerns and there are "too many things going on". In that way the message is muddled, and the consensus necessary to identify the meaning is lacking. But the message can also be muddled if the artist is deliberately self-contradicting as to avoid commitment.[14] In this case, too, intent can be impossible to read, and the meaning fails to materialize. Contrariwise, if the artist has clearly honed the relevant point of view and conveyed it honestly, the viewer will not only understand both the intent and thus the meaning of the artwork, but will concomitantly be able to experience trust in the artist and thus absorb an additional viewpoint. This is a kind of knowledge-acquisition as a new view has, with clarity, been provided; and honesty had an important role in this in that it communicated the intentions.

[13] Bruno Latour, "Visualisation and Cognition: Drawing Things Together" in H. Kuklick (ed.) Knowledge and Society Studies in the Sociology of Culture Past and Present, Jai Press vol. 6, p. 21.

[14] This sometimes happens when artists (or their dealers) seek refuge in the camouflage of "ambiguity", which is often done for monetary reasons as it is easier to sell one item dressed under various cloaks.

5.3.4 Belief

> Doubt is an uneasy and dissatisfied state from which we struggle to free ourselves and pass into the state of belief; while the latter is a calm and satisfactory state which we do not wish to avoid, or to change to a belief in anything else (Peirce, 1986).

In the previous sub-section, I stated that trust is the bridge to belief. This point is obvious, but let me expand on that by briefly quoting the philosopher Sarah Moss. She states, "Suppose you find yourself disagreeing with an epistemic peer—namely, someone whom you antecedently trust just as much as yourself to form true beliefs about the question under dispute. How should you update your own beliefs upon learning what your peer believes?" (Moss, 2018, 160). She follows this by giving an example of differing beliefs between a pilot, who thinks there is enough fuel, and a co-pilot, who does not; how much does the pilot change her credence of belief after hearing the co-pilot's views? The pilot will be affected, no doubt, by the co-pilot's belief; perhaps the pilot went from near certainty to much less certainty after hearing the co-pilot's belief. Surely, we have all felt the strength of our beliefs shift in similar situations.

This example, taken from her book *Probabilistic Knowledge*, is a piece of a larger argument wherein Moss shows that belief is not to be seen in terms of propositions (with its necessary entailment of full beliefs, its structured contents, and corresponding rules for truth satisfaction), but instead in terms of credences or probability functions. She argues that the paradigm for belief should be modeled on the fact that we very often hold beliefs in degrees e.g., in credences—a 0.6 belief in something or a 0.4 belief—instead of full belief. In her example above, perhaps the pilot went from 0.9 to a 0.6 credence. In this, Moss argues that our credences of belief are often shifting, and she argues this within a probabilistic framework that is concomitantly arguing against a propositional basis for knowledge.[15]

Belief events then become context-driven, and we not only are influenced by other people but our beliefs are added to those of other people. Moss states it thus, "According to theories in the tradition of GRICE 1967, the contents that we assert are added to the common ground of our conversation. According to this model of conversation as developed by STALNAKER 1978, the contents of assertion are sets of worlds…" (Moss, 2018, 29) (capitalization in original).

How we come to belief is currently a much-researched topic, and much of this research is often focused on social issues. This is an obvious locale for the subject, for how we come to belief when faced with a context of completely inanimate objects (such as was the paradigm example in traditional empiricist philosophy) and how we come to believe something when interreacting with humans are two distinct situations. To quote another contemporary philosopher, Rima Basu, on the importance of this distinction, "To begin with, we don't relate to people in the same way

---

[15] See Moss (2018): "…my argument for a fairly radical conclusion, namely that probabilistic contents should ultimately replace propositions as the fundamental contents of belief" (Moss, 2018, 14).

that we relate to objects" (Basu, 2019). She then quotes Rae Langton: "'we don't simply observe people as we might observe planets, we don't simply treat them as things to be sought out when they can be of use to us, and avoid when they are a nuisance. We are, as [the British philosopher P.F.] Strawson says, involved'" (ibid.).

Because each individual confronts a sandstorm of data, so to speak, such that our absorption of the world is neither straight-forward nor is it easily systematized into a formulaic structure, we do not absorb everything that comes at us nor is our consciousness a mirror reflection of that external data; thus the data is filtered. Some of the multitude of sensory data or qualia makes it into our consciousness and some remains nonconscious, and some doesn't get processed at all.

In this sorting, the views of others are significant sources of influence. They change the weighted value of what gets noticed,[16] and encourage us to notice things we hadn't before. This is because other people are significant in singular ways.

And art is an epistemic substitute for people. When I look at an artwork, I am looking at another's first person experience, and thus, indirectly, looking at them. Art is a way we have of indirectly, (and thus safely) interacting with others in order to empathetically understand them and thus ourselves, and therefore to be able to corporately engage with a social world.

Therefore, my argument is as follows. I propose a view of aesthetics that links the act of perceiving an artwork to belief formation, and thus that counts the (visual) art object as evidence for belief in the referent of that art object.[17] The referent, properly understood, is a changing entity. It is not a rigid designator. Initially (when the art is still in the studio), the art object refers solely and exclusively to the first person experience of the artist. Then, when the art object is seen by an initial viewer, the viewer—through experiencing the low-level (qualia) of the artwork and its affordances—experiences thus the first person experience of the artist as instantiated in the artwork.[18] And the viewer's first person experience in then conjoined with the embedded reference to the first person experience of the artist; the aesthetic experience now is constituted by both first person experiences i.e., their conjoined qualia.

---

[16] Similarly to neural nets.

[17] There are additional issues here cannot be fully treated but deserve notation. The standard way to look at belief is intentional: To stand in a relation of belief is a relationship between the bearer of the belief and a proposition. But let us discard this propositional view and think more broadly. Do we believe in the referent or in the meaning of a term? What is belief tied to? For example, "I believe in the president." Is the belief attached to the referent of "the president"? In this, the words "the president" refer to the person who is the president; it is simple ostension. Meaning though is always a bit harder to lasso, and the meaning of the sentence is distinct from the meaning attached to the terms "in" or "the president". Meaning is though derived from conjunction, and must be thought of additive along credence values. Thus, belief is attached to the meaning through the locale of the referent.

[18] The art object refers solely to the first person experience of the artist in those situations when it is in the state immediately after production and before the artwork has had significant exposure to the public. It is the state when it is looked at by the first viewers.

The referent has now slightly shifted for that specific viewer as the artwork is referring to the conjoined qualia e.g., my first person experience of JoseyNewArtist (which entails the artist's first person experience). As the art object becomes discussed by the public, the conversations around the art object engender interpretative constraints, and what the artwork refers to thus shifts again. The interpretation of the artist's first person experience has been altered by the interpretative constraints e.g., this is everyone's experience of JoseyNewArtist. Thus, the art object now refers to something slightly different, and the meaning of the public art object is that that encapsulates that first person experience of the artist combined with the first person experiences of the multiple viewers. (The social aspect of this is explained more fully in Sect. 5.4.1.)

In this, clearly, the referent of the art object is an abstract object. To slightly rephrase the process. There are two different contexts for this: (1) an artwork by an artist who has little or no public following, and (2) an artwork by an artist that has a public following and a consensus view regarding that artist. In the first instance, the referent of the art object is the artist's first person experience and thus derivatively the artist. This has been discussed. In the second context, it is also in this case that the referent of the art object is the artist's first person experience and thus derivatively the artist, but it is slightly different as public conversation has entered in as a variable. In the later stages, as the art object gains viewers and an interpretation congeals as the consequence of the conversations around the art object, the referent of the (new) viewer's belief then incorporates this conversation.

Like Moss' pilot, we are influenced. Thus, when a new viewer stands in front of an artwork that has a place in history, which is an interpretive agreement about the artist's first person experience that is communicated via the artwork, then that new viewer sees that artwork *in terms of* the agreed upon meaning of the artwork. For example, when I look at a Rembrandt I experience it not only directly as I would an artist whose work has not been publicly seen, but I experience also the consensus views of others.

Thus, aesthetics is the act of perceiving an artwork that links that visual experience to belief formation. In this, aesthetics counts the (visual) art object as evidence for belief in the referent of that art object. The referent of the artwork is an abstract object, construed as a non-a priori, non-inert, and non-eternal object, and is an example of a mind-dependent ontology and a constructivist epistemology.

Meaning is thus derived from conjunction, involving both the first person experience of the artist, the first person experience of the particular viewer, and the interpretive views of others, and must be thought of additive along credence values. Belief is attached to meaning through the locale of the referent. It is an analysis of the heuristics of the aesthetic experience as an instance of belief and of the art object itself as evidence for that belief.

This aesthetics rests on a philosophy of mind that, like Moss', rejects propositions as the paradigms for the contents of beliefs. Instead, I would argue that our beliefs often have as their referents points of view, and those points of view are tied, at least partially, to people. This is why a probabilistic function is the correct way to analyze them. There are not abstract propositions, they are not "what" statements, they do not solely involve the function of semantic memory and are thus not about thus about the naming relation between an object and its predicates. Instead, they involve values, as well a complex and inter-dependent parts. We believe in things, in people, in points of view; we don't just believe "that" statements.

My point of view runs contrary to much in aesthetics. For I would argue that what constitutes the art object is not the mere artifact itself. Rather, the art object, as an ontological entity, is an abstract object and it changes over time and depends largely on the conversations around that object. The physical art object is the sign that points us to the referent, and then is evidence for our belief in that referent. For example, if I were a person in the late 1600s and I was presented with a Rembrandt, I might be struck by the way my reactions (partially based in my biology) to his treatment of illumination—the lights and darks would be noticed and would affect my first person experience. In addition, I would understand that that treatment of illumination was also a representation of his qualia—the way the world was experienced by him. I would then interpret that painting as the way he saw the world. This part is uncontentious. Artists give us their version of the world. I would easily have said, "So this is the way Mr. Rembrandt sees things."

There is, though, a second step in the creation of meaning, and it is thus: Conversations. It formulates the general consensus: "This is what *we all* think Rembrandt's first person experience was." While the first viewers looking at the artwork will experience the referent of the artwork as the artist's first person experience of the artist, the referent doesn't stagnate. It is not a rigid designator. For, as an artist's work gains more public viewing and more public consensus, there is an additional process of adding in the viewers' interpretations of the artist's first person experience. So when I, as a new viewer in the twenty-first century, come to look at Rembrandt, I am not only responding to the qualia of the illumination (both my first person experience and by inference Rembrandt's first person experience), but am responding to the prior social consensus of what that illumination means: "Rembrandt seamlessly melded the earthly and spiritual."[19] Meaning is thus accrued. What the viewers of today see when they see Rembrandt is the layered meaning that has accrued over the centuries; thus, we see it in terms of meta-earthly transcendence. We are like Moss' pilot example: Our views have been influenced by others. This meaning—e.g., what the artwork itself means—develops over time, and it is that abstract object that is created and it is that abstract object that belief then adheres to. (More is said about this in Sect. 5.4.1.)

This point of view relies on a notion of belief that is a dynamic system existing within the context of embodied experience, which itself relies on inductive practices and pattern recognition as well as the universal constraints imposed by biology. The latter, of course, involve the neurological constraints of vision including both affordances as well as saliency maps. Such a dynamic system of belief formation would not adhere to the rigid structure of linguistic grammar, and should then be more

[19] This is a quote from the Wikipedia article on Rembrandt, or what counts, seemingly, as the world's biggest consensus statement.

properly modeled on something like probability functions. The latter allows for interpretative flexibility as a probability distribution shows all the possible values that a random variable can take within a given range. Bounded sets play an important role in this, as the range will be bounded by the minimum (e.g., Rembrandt's illumination is just realistic lighting with no meaning or affordances attached—"no big deal!") and the maximum (e.g., Rembrandt's illumination is a sign of transcendent God-light—"earth-shattering!"). This is why sets are essential in the analysis and why grammatical propositions are an inaccurate way to model belief.

A quick but pertinent reiteration and then an additional comment should be made. The reiteration: Since the referent of the artwork is the first person account of the artist as experienced by the viewer(s), then aesthetics is the activity of attaching belief to the meaning inherent in understanding that referent. Now for the additional comment. Thus, aesthetics cannot be understood through the semantic description of the artwork itself. I can describe the art object, whether it be a Whiteread or a Monet or a Rembrandt; I can detail the colors and the shapes, and I can even translate some of that into words that describe the piece's general metaphorical meaning. For example, I can posit that the Whiteread piece means sadness or loss, or something like that.

But to articulate what my bodily and mental *experience itself* was—what it felt like to stand in front of it and go from the state of not-knowing the piece to the state of knowing it, to put into words the intense feeling of pleasure when that happened and when I appreciated and "fell for" what was being told to me—that is a very different matter. It is a bit like "falling" for a person perhaps; that's why people say they "love" a piece of art or a piece of music with the same intensity. And in a similar fashion, one might be able to describe the virtues of one's beloved by detailing their character or physical traits, but one would be hard pressed to describe the love itself. Describing the physical artifact and explaining the experience are two entirely different things.

The task of aesthetics is the latter. It is the experience we have when we are standing in front of an artwork, *and* the knowledge that is gained from understanding the referent of that object—this is aesthetics. Experiencing an artwork is then one instance of gaining access to something—that is, knowing something—that is a kind of relationship one has to the world. It is a way to belief.

And clearly it is more complex than what a that-proposition can encompass. The view that limits knowledge to the realm of propositional statements limits then knowledge to that small range of experience that can be formulated as a "that" statement. Clearly, limiting one's epistemology to such narrow requirements means fundamentally not addressing the myriad other ways that we dispel doubt, which is the very essence of claims to knowledge. To refer again to the opening quote by C.S. Pierce, "Doubt is an uneasy and dissatisfied state from which we struggle to free ourselves and pass into the state of belief…" The purpose of the epistemological enterprise is to provide an answer to how it is that we go from doubt to belief, and as we travel that chasm in various ways and in a myriad of contexts it is imperative that our theory is able to address them all.

This is my fundamental criticism of the aesthetics of Nelson Goodman. All he offered was a semantic account of the artwork *after* one had experienced it. This account was of the concrete object in the world referred to by the artwork with the metaphorical definition attached e.g., "Before me is a picture of trees and cliffs by the sea, painted in dull grays, and expressing great sadness" (Goodman, 1976, 5). This allowed him to be consistent with his nominalist account that forbid sets in mathematics and abstract entities in philosophy, and thus would not have allowed him to countenance a notion of either probabilistic knowledge or possible worlds.

But, in fact, there are sets in mathematics. And I would argue as well that there are abstract entities in philosophy. More importantly, perhaps for the purposes at hand, there are first person, non-extensionalist accounts in our experiences of looking at art. And those experiences aren't remotely captured by a semantic, referential, and extensionalist analysis such as his quoted statement given above. Furthermore, while Goodman did allow a role for emotions, it was not clear how they remained within his nominalist account and did not fall into his much-feared platonism.[20] And while Goodman did use qualia, his definition (i.e., as concrete particulars at a particular place, a particular time, a particular color), is very different from how qualia is usually defined today i.e., "…consciously accessible features that are non-representational and that are solely responsible for their phenomenal character."[21] The debate here hinges on just how to define qualia, but if consensus were to converge on this latter and newer definition, then this, too, would be a problem for Goodman. For while he was eager to keep his qualia within a nominalist account and as the atomistic constituents of reality, his qualia did though carry information about external objects because qualia was constituent of concatenated and complex states of ordinary objects and thus then required (indirectly) to represent. Relatedly, only reference was allowed—meaning was forbidden.[22] To quote Evans' view of Frege: "This position seems barely intelligible."

As Goodman's views were solidly founded upon his nominalism and upon his semantic account of reference, this dovetailed nicely with those twentieth century theories that held language to be an isomorphic reflection of experience, and thus led to relating belief to language. While this twentieth century view has been more recently contested, it seems at first harmless when traced back through its philosophical moves. It is worth looking at this slide from plausible assumption to difficult conclusion.

The theory was this. Beliefs have their foundation, largely, in perceptual experiences, and perceptual experiences are related in language, which has as its twin:

[20] See: Shottenkirk 2009, footnotes on pages 37, 146, 149, 160.

[21] See: Tye, Michael, "Qualia", *The Stanford Encyclopedia of Philosophy* (Fall 2021 Edition), Edward N. Zalta (ed.). https://plato.stanford.edu/archives/fall2021/entries/qualia/

[22] My position is the opposite of Goodman's for I agree with today's consensus opinion that qualia don't represent, since in my theory it is the artwork that represents the qualia. If the qualia were to encode information about external objects they would be said to be representing those objects. But in my theory the transfer of the qualia from the artist to the viewer is a qualia to qualia transfer, and does not involve representation.

Propositions. It is important to remember the layers of commitment for these traditional, now contested views. As Frege and Russell thought grammatically structured sentences had their meanings in propositions, and were thus bearers of truth through their relationship to the similarly structured—though ontologically different—propositions, so many have argued that our perception has its contents in our grammatical representation of that experience, which in turn of course has its structure similarly concatenated in the abstract object called "propositions". And, once again, it in turn has its constitutive foundation in concepts. Predicates apply to objects because those particular predicates instantiate a concept that is appropriately applied to the object. To say that perceptual experience is structured by concepts is to say that an understanding of perceptual experience is ultimately traceable back to concepts. Thus, "conceptualism", as a term used within an analysis of perceptions, argues that for any perceptual experience, that experience has as its content a Fregean proposition.

But this traditional view has been challenged, with some directly stating their opposition to identifying the contents and structure of beliefs as the same as the structure of the sentences that express those beliefs.[23] Thus, I would also argue against also the standard picture that holds that belief is a relationship between the bearer of the belief and a proposition, which also incriminates a notion of concepts that exists antecedent to our distinct, individual, subjective and embodied experience.

Instead, I locate belief within the realm of the embodied, communicative experience. Art is the paradigmatic embodied experience; one that incorporates the views of others while developing empathy for individuals. In this, art is the paradigm of collective behavior.

5.4 Art as Conversation

5.4.1 Social Cognition

When we stand in front of an artwork and say, "yes, I like it", what are we saying yes *to*? Let me phrase this slightly differently. At the very uncontentious minimum, when we assert that we like an artwork we are assenting to a mode of presentation. We have a belief in, a trust in, the validity of that mode of presentation. Therefore, our assertion points to a belief *in what*, precisely?

The last sub-chapter provisionally answered this question i.e., we consent to the point of view conveyed. This latter is a conjunction of the artist's first person experience as symbolically represented in the painting and, as well as, the consensus of the viewers' assessments. But I want here to expand on that, and tie it more fully to

[23] Crane, Tim. 2016. "Lecture: Unconscious intentionality" II International Conference on Philosophy of Mind: Thought & Perception. University of Minho, School of Psychology, Braga, Portugal.

the notion of social facts, explicating as well the philosophical commitments that underlie the theory of social facts as well as some other related and relevant philosophical perspectives that address social cognition in general.

On Social Facts was published by Margaret Gilbert in 1992 and significantly furthered the notion of cognition that reached outside the traditional domain of the individual. Building on the phrase "social facts" that was introduced by Durkheim, and influenced by Wittgenstein's private language argument, she addresses the related notion of collective belief in the following:

> …collective belief: roughly, a group belief is a jointly accepted view. That is, it is a view that each of a set of persons has shown willingness to accept jointly with the others. A jointly accepted view is not to be confused with a view that all or most of the people concerned personally accept. One may accept a view jointly with certain others but not believe it personally.
>
> That a certain view is jointly accepted by certain persons is an extremely important type of fact (Gilbert, 1992, 20).

Gilbert examines the views of the social theorists such Max Weber and Émile Durkheim, among others, in order to argue that groups are collective agents whose views cannot be reduced to just the sum total of the individuals. Instead, this membership involves a commitment to group action (Rosen, 1989). Thus, Gilbert's notion of social facts contends that we all hold beliefs together that cause us to act and believe together. Giving credit in her *On Social Facts* introduction to Charles Taylor's skepticism regarding David Lewis' game theory notion of "common knowledge" (wherein Lewis maintained that common knowledge only meant that everyone had a reason to believe a certain proposition)[24] she develops Taylor's criticism in a 2007 paper and states:

> The writer-reader relationship, incidentally, would seem to be a version of the situation on which Taylor focuses. The writer purports, implicitly or explicitly, to "address" the reader with his words. The reader is supposed to "get" what he is saying: but more than this…The flavor of such locutions is, I take it, to invoke something more like the creation of a common focus than the transmission of information from one mind to another…Or so I say to whoever reads this, conjuring the creation of a new public space, hoping that by reading what I write, he or she will, as far as is possible, join me in that space (Gilbert, 2007, 6 footnote).

The claim here is that "common focus" is a way we have of bringing each other into a space—conceived of as a constructed ontologically existent entity—that is jointly shared by us, which in turn creates social facts. It is not restricted to the transference of information; rather, it is more than that: It is "the creation of a common focus"—i.e., a way of looking at the world, a shared perspective.

This point of view historically follows the development of social epistemology, a movement that began in the 1960s and 1970s that included an examination of cultural biases, group beliefs, and the constructivist nature of facts. This movement, seen primarily in analytic philosophy countries and championed most prominently by Alvin Goldman (Goldman, 1999), dovetailed with some instances of continental philosophy done by the deconstructionalists and post-structuralists (e.g., Foucault,

[24] See Lewis, 2008.

Barthes, etc.) with interests in "how knowledge operates as a principle of social organization" (Fuller, 2002, x). The unified point shared between these two philosophical traditions was not only that history and culture mattered, not only that we are influenced by each other's views but—more importantly—that we form views jointly. To state it more bluntly: Belief is not just held individualistically; cognition telescopes beyond the cell wall.

This goes very far in explaining what it is we assent to when we say we like a work of art. But firstly, the denials: Our belief in the art experience is *not* directed to the artifact itself, because the artifact is not an inanimate object in the way a rock is—the artwork is derivative of a human and therefore is social; in addition, our belief is *not* directed to objects in the world via a depiction function, because, for one reason, we already have that sort of belief established and wouldn't need art for that. Instead, the answer begins with knowing that saying "yes" to an artwork not only hinges on viewing the aesthetic experience as a part of epistemology and seeing our understanding of art as a kind of cognition, but also in agreeing, at least significantly, to view the artwork from the point of view of the group.

This may sound like an odd point, but it is, I believe, central to the experience of art. We are so conditioned to think that our ideas are ours own alone that we are unaware of those many instances where we are looking out at the world from within the framework of a group. For example, when I look at Rembrandt, I am looking through the scrim of a social belief regarding what Rembrandt is saying. Thus, our saying "yes" when we stand in front of Rembrandt signals a belief in a social fact.[25]

This latter obtains as a consequence of the fact that our visual experience of the world is not a passive act; this fact causes us therefore to edit the world around us both consciously and non-consciously. We choose; and then we construct. (And the choosing is itself dependent upon trusting.) This point was primarily driven home by the midcentury work of Quine who argued that the world does not come pre-chunked, and "no statement is immune to revision" (Quine, 1980, 43). Science itself was the initial target for this, and Quine developed the influential view that science is revisable; or to brashly paraphrase—science is more like Heraclitus' river than it is like facts engraved in stone. Not only do we sometimes have new data but we choose which data to pay attention to, and we decide how to frame and interpret that data. Thus, theory changes, too.

This ability of ours to edit, sort, and categorize the uncountable number of discrete bits of information is the basic mechanism of our epistemological experience, and the ontological reality underlying that epistemology is a world that not only does not come pre-chunked, but a world we are constantly remaking. Thus, for

[25] The objection might be made that this paradigm would not apply to any artworks that do not have a large following or in fact no following at all. I have tried to make the point that in those cases, the viewer is attempting to understand the artwork more on the basis of parsing simply the first person experience of the artist that is symbolized in the low-level P-unitates of the work e.g., "the silky brushstrokes of yellow here fade off into delicate textures of pinks and greys; I understand this to mean a kind of gentleness". But, importantly, even this private interpretative move is done on the backs of endless amounts of social facts; we have agreed to all the symbols of these things.

Quine, science was constantly being revised, as consequently were our views of reality. The divide between objective and subjective were likewise shunted aside and replaced with the adoption of the term "intersubjectivity": We know things together on the basis of each of our subjective experiences (this is the beginning of our awareness of the world), but then in consort with each other we merge together our subjective assessments and it is in the set union of those assessments wherein we agree on the fact of the matter. Thus, entities are not pre-structured prior to our usage, and discovering takes a back seat to constructing. This freedom to construct is also found in Gilbert, who supports Taylor's position that sees social spaces as ontologically robust: "In his conception, then, public space is constructed, not discovered" (Gilbert, 2007, 7).

My theory is in consort with this; our formulations of reality are part of a dynamic system and temporal change in an inherent part of that. In addition, intersubjectivity is central to my epistemology. This is especially true with information gained from sensorial experiences that entail the visual system. It is interesting to note that Quine's philosophy was not based on a non-sensory or ideal mode such as that posited by such foundationalists as Descartes, but rather based on sensory inputs as bodily facts and tactile realities e.g., "neural inputs". These included such things as the air striking my skin, or visual realities such as the light waves striking my retinas, etc. These inputs of the individual were then fodder for the social process of science, which he formulated on a pragmatism similarly to Pierce's (Shottenkirk, 2009).

And pragmatism prevails here in my theory as well. It is the artist's choice of the subset of data that is the beginning of the process toward the creation of an artwork. When the artist chooses a subset of the data, and presents that subset within a particular attitudinal framework that references the embodied first person experience itself, the artist gives over to the viewer a translation (transduction) of his or her first person experience. The work then enters the world, and though at this point that art object has as its referent the artist's original subset of data, which is properly viewed as the cause of the artwork, it also is in the process of additionally gaining something else, too. That additional constituent is the views of others. This becomes the attitudinal framework with which the artwork is now viewed.

It is here where the pragmatism enters in its fullest role. Consensus constructs the social facts and those are accepted on pragmatic grounds. Thus, the referent of the artwork is the socially agreed upon view of the work, e.g., its meaning. Therefore, when we say "this is what that work is about" that content is the end product of a long process of consensus building and is constituent of the epistemological basis that forms public language.

For example, let us take a piece by Agnes Martin. This painting, "Friendship" (1964) is representative of much of her work.[26] Our first gist of the painting will convey much of its tenor—calm, rigorous, orderly. This information—these affor-

[26] To view a color version of the painting along with an essay regarding it, see: https://openheart-project.com/the-path-post/agnes-martin-friendship-1963/

dances—would be there for us within the first few millisecond. And while this would be a beginning for us, it wouldn't suffice as the end. More knowledge would be needed in order to fully appreciate it. And while we might now, in the year 2023, be able to discuss this piece, what an Agnes Martin "means" was not (as it is not with artists in general) there in the first instance of her practice. It takes a while because the meaning is not fully under the control of the artist herself. That is both due to the development of conversations around the work, (where the meaning is, in part, contributed by the audience) but also because it necessitates the development of the artist's craft and style, honed from teachers' critiques and a developed sensitivity to the zeitgeist of the culture. And then, in a cyclical manner, by the artist who in response to the audience's reports regarding what was being communicated, alters or changes the work itself. Martin's work has been said to have taken a rather longer time to congeal into a delivery packet of content; the New York Times critic Holland Cotter wrote that it took her 45 years, and he describes the meaning of her work in the following way:

> View her paintings from several feet away, and their surfaces—whitish, pinkish, grayish, brownish—look hazily blank, as if they needed a dusting or a buffing. Move closer, and complicated, eye-tricking, self-erasing textures come in and out of focus. Move in very close, and you find that the surfaces are marked with hand-drawn lines, often faint but always firm, and regularly spaced, like the lines of a musical staff, or an accounting ledger, or a school notebook… Her art—controlled, quiet, tough, painstakingly poised…[27]

Mr. Cotter did not make this up on the fly; this point of view was already more or less the agreed-upon consensus of Martin's work although he further refined, articulated, and codified the view. This review was one more step in the process of the developing, morphing abstract object we would call "the meaning of Agnes Martin's work".

How does a socially constructed fact in the artworld get established? The easy answer is through critical evaluation, repeated assertions, and monetary value. The harder answer is a philosophical one. When Descartes wondered at the failure of his ability to discern whether he was asleep or not or whether he was mad or not, his problem was grounded in an individualistic notion of knowledge. But, if he had thought in terms of social facts, or in common knowledge and the epistemic satisfactions for truth conditions that were grounded in that ontological entity we call group knowledge, he could have turned to those around him and said, "hey, do you think I'm crazy?" Had he done so, his epistemic process would have been profoundly different! Not to mention the history of philosophy…

[27] Holland Cotter, "The Joy of Reading Between Agnes Martin's Lines." *NYTimes* Oct. 6, 2016. https://www.nytimes.com/2016/10/07/arts/design/the-joy-of-reading-between-agnes-martins-lines.html

5.4.2 *The Group*

Thus, it is the notions of social facts and common knowledge that give rise to a philosophical perspective that allows us to make sense of larger phenomena such as the comprehension of artworks. Said such, it seems certainly plausible and well on the road to uncontentious. But one is immediately led to more prying questions. What exactly are the psychological processes that allow us to come to agreement on points of view and how do we analyze those philosophically?

With this in mind, it would be worthwhile to focus more specifically and with a closer frame in order to examine the specific mechanisms of empathy, filial alliances, and actual love that allow us to pull together and see things from a joint perspective, which in turn leads us to support one another and build productive social structures.[28] Likewise, it would be worthwhile to focus more specifically on those instances of peer pressure, on the fear of social ostracizing, and on the mammalian-brain willingness to adopt status-accruing beliefs that encourage us in prejudicial attitudes that in turn lead us (as organized groups) down the road of cruel triumph over those who might be in a weaker position than we.[29] Worthwhile goals, surely, but largely beyond the confines of this book.

We can, though, go far in explicating a few of the basic mechanisms that allow us share perspectives in those situations that yield the culturally-influenced practice of aesthetics. I will therefore address the following topics: Joint attention (in psychology), and testimony (in philosophy), First, joint attention.

5.4.2.1 Joint Attention

The development in psychology of the theory of joint attention began in the 1970s with the introduction of the term by Jerome Bruner. The traditional paradigm for this is the interaction of the mother and baby or small child. The mother points to something, expresses some thoughts about it or attitude toward it, and the small child responds. Anyone who has had a baby or even been around small children knows this phenomena; it seems to be hard-wired into our species (as well as with some other primates, and perhaps horses as well). This begins as early as 2 months, but by 18 months babies are well versed in this process and can communicate with those around them despite the fact that language is only in the rudimentary stages.

The discipline divides joint attention into three sequential states—the triadic, dyadic, and shared gaze. The latter is the first stage and is seen in very young infants, while the former is thought to signify full understanding (Legerstee et al., 2007). Thus, triadic attention is the stage at which joint attention is fully developed and is

[28] See: Liz Camp & Carolina Flores, "That's all you really are: centering social identities without essentialist beliefs" in *Mind, Language, and Social Hierarchy*, ed. Sally Haslanger (Oxford: OUP), (forthcoming).

[29] See: Mills, 2007; Antony, 2020.

seen when the baby looks back to the adult (e.g., mother) after looking at the thing referenced. Again, this kind of exchange would be familiar to anyone who has had or been around other people who have had a baby, and they would have seen the baby checking the mother's facial expression to see her evaluation of the referenced object. This is a kind of confirmation stage where the baby is understanding not just what is being referenced but *how* it is being referenced and therefore the attitude to hold toward that thing. (Note, this paradigmatic relation is like art.) For example, "Don't touch the stove it's hot!!" This is a very communicative exchange. The in-between state, called "dyadic" is when there is a conversation-like behavior that both adult and baby engage in, such as "let's pick up the ball".

While the general description of joint attention is more or less agreed upon, the modeling of it is not. The general description runs something like this: "An individual is *monitoring attention* when she takes a third-person, observer's perspective on a second individual, and attends to what the other is attending to" (italics in original) (Siposova & Carpenter, 2019, 261). But the psychological and philosophical explanation for this definition is contended, with a few competing paradigms for this. For some, joint attention signals intentions: The exchanges between the infants and other people are "in terms of concrete intentions rather than thoughts and beliefs" (Tomasello, 2014).

But this perspective does not hold for all. Thus, for others, the reference is not to an inner mental state. An example of this is the researchers Racine and Carpendale who note the philosophical underpinnings of this research in psychology: "The scientific study of how infants develop social understanding necessarily, but not obviously, involves philosophical claims about just what social understanding is" (Racine & Carpendale, 2007, 4). They state their case against the role of intentions in the following:

> What we would like to stress is not simply that an intention would not be clear to an interlocutor, but rather that there is no intention here… What are often considered to be 'mental states' do not exist as inner entities that can be referred to. And if infants and children could not be aware of such states in an introspective sense, they could not label such states or apply such an understanding in order to reason by analogy about others as is entailed in simulation theories of infant social development. How, then, do children come to learn to understand others in psychological terms?…However, capacities to attend, want and exercise will are in fact preconditions for the operation of what Wittgenstein called the language-games of attention, desire and intention. Language-games are tied up with the activities and natural reactions that they describe (Racine & Carpendale, 2007, 6).

After discussing Wittgenstein's arguments concerning the difference between saying and showing, the researchers state, "Instead, people use language to convey meaning, and this is always within a sequence of interaction" (ibid.). I would agree with this.

Intentions do not properly frame the issue. While it seems true that infants understand not only the bare reference of the eye gazing or finger pointing but also the attitudinal content that is encapsulated in the liking or not liking that the adult has expressed, what is less than conclusive is how that attitudinal content is conveyed. To say it is consequent to "understanding intentionality" is to say too much. For the baby or small child is in general trying to please the adult—this is a main motive of infants—and thus will mimic the adult's viewpoint, which is automatically adopted by the baby. (One has only to think of the biological and survival role of mirror neurons.) Clearly, the mimicking of behavior and attitude could be accomplished without any peering into the mental contents of the adult's mind. Instead, all we need here is a joint attention to affordances.

For example, the parent acts afraid in front of the stove and tentatively puts her hand near it, screams in horror and pulls it away crying "hot". The baby understands the negative valence in that reaction, especially after the parent has put the baby's hand close enough to feel the heat. (But of course not burning it!) There is a shared physical and neural reaction to the heat and thus the attitudinal framework is reinforced. The same process holds true with such pleasant things as playing with a ball or licking an ice cream cone. The infant will generally adopt the attitude of the parent toward the object. As infants are extremely interested in interacting with other adults and children, they will turn away from other things in order to engage in joint attention (Carpenter, 2010). Infants want to interact, want attention, and want to please because they want to be loved and taken care of. "Imitation is the highest form of flattery" is true for infants, too. This allows babies to learn which things are to be avoided and which things are to be desired. Affordances have a biological basis that doesn't require the mystery of other minds.

5.4.2.2 Testimony

As a fruitful source of recent philosophy, testimony and its related notion of evidence have been researched by epistemologists who recognize in both of these topics some of the most difficult problems in philosophy, and the keystones to the discipline itself. For without an account of testimony we won't know when to count testimony as evidence, if we don't have a coherent account of what counts as evidence then accounts regarding belief cannot be established, and without belief we cannot get any account at all of knowledge. Thus, the interest in testimony and evidence. One thinks most immediately of evidence in a situation like the scene of a crime—the gun, the DNA, etc. These bits of evidence point to the perpetrator, and that seems easy enough. The testimony would be that of the witness or co-conspirator. The two terms "testimony" and "evidence" though are sometimes intwined in our language: The witness "gives evidence" against the defendant, the gun itself is "testimony against the accused". But if we try to philosophically pry them apart, what can be said about other kinds of testimony, and what counts as testimony?

Jennifer Lackey and Ernest Sosa, in the introduction to their influential 2006 book titled *The Epistemology of Testimony* asked, "are the normative requirements for testimonial justification and knowledge a burden that falls primarily on individual knowers, or are they largely satisfied by features of the social interaction that takes place between speakers and hearers?" (Lackey & Sosa, 2006, 2). In this edited volume the included papers consider evidence's role in such traditional problems as justified true belief, reductionists versus non-reductionist accounts, etc. But in the introduction, after noting that some of the authors have a wider view of what counts as testimony than others and using Elizabeth Friker's standard of "telling generally" (which would include aesthetic matters) as an example of this, the editors then make a significant disclaimer as to what counts as knowledge:

> But not every expression of thought is appropriately regarded as an instance of testimony. For instance, suppose that we are walking down the street and I say, 'Ah it is indeed a beautiful day'. Suppose further that such a statement, though it expresses my thought that it is indeed a beautiful day, is neither offered nor taken as conveying information for any proposition; it is simply a conversation filler, comparable to a sign of contentedness. Or consider a case in which Ned tells a joke…Since the concept of testimony is intimately connected with the notion of conveying information, either in terms of being offered as such or as being taken as such, it is arguable that both examples should fail to qualify as instances of testimony (ibid.).

I would argue that this position contravenes the argument presented by Margaret Gilbert in Sect. 5.4.1 when she was arguing for Charles Taylor. I will repeat the quote: "The reader is supposed to 'get' what he is saying: but more than this…The flavor of such locutions is, I take it, to invoke something more like the creation of a common focus than the transmission of information from one mind to another" (Gilbert, 2007, 6 footnote).

I would vote with Gilbert here. The core mission for epistemology is to explain how we go from doubt to belief. That process—or that mental and bodily journey—is accomplished in many myriad kinds of instances including the ascertaining of scientific claims (however open to revision they might necessarily be), to adopting the narrative of a historical event, to conveying a "common focus" in a social context, to concluding that a certain artwork is meaningful and successful. In all these situations, plus innumerable others, the subject (or thinker, or one experiencing the phenomenon, etc.) has managed his or her way through the bramble of doubt and come into the calming state of belief. It is epistemology's job to explain that journey.

Thus, I would argue that the comments of a trusted peer who gives us evidence for an attitude toward something or evidence for a framework of values, is, in fact, providing testimony. When I say, to use the editors' example, "Ah, it is indeed a beautiful day" to my companion, I am giving evidence for an attitude of gratitude, for a positive framework of values. I am giving testimony that supports the claim that we ought to be grateful. I am giving testimony that supports the evidence that it is, indeed, not only a beautiful day but that that awareness of the beautiful day is a constituent of a larger awareness of the wonder of the world. In this, my testimony is pointing to this larger fact, it is substantiating this larger fact, it is providing evidence for this larger fact. To refer back to Gilbert again, it is the "creation of a common focus" rather than "the transmission of information", and it leads the hearer to "join me in that space".

Some recent theorists have joined in the push to cut philosophy's traditional ties to information defined in purely informational, linguistic, or propositional terms and investigated testimony more widely. Considering the distinction between information and knowledge, wherein the latter is both broader and different, Federica

Malfatti argues testimony can bring to bear a new point of view in which "grasping a fact means, at least in part, to be able to properly embed the corresponding informational unit into one's web of cognitive attitudes" (Malfatti, 2021, 1354). In other words, knowledge is not only broader than that of information, but instances of testimony that broadens knowledge includes "a process of adjustment and rearrangement" in one's cognitive attitudes (ibid.). I take this to mean something like Gilbert's "common focus".

Other philosophers have also articulated a broadened notion of testimony. For example, artists give testimony to violent events, and this therefore demands a consideration of such things as the role of cognitive attitudes in the reception of those artistic works that recount violent events and the artists' role as witnesses, which in turn rely on the audience's trust in the artists' testimony (Jones, 2019). Let me give another example that deals directly with the transference of the artist's first person experience. While philosophy has historically claimed that that the first person account is irremediably private, the philosopher Yuri Cath has pointed to the contrary public fact of artworks conveying the "what it's like" of various experiences (Cath, 2019). Cath gives the example of the plethora of articles, YouTube videos, books, and movies that purport to give us knowledge of what it is like to go to war, have a baby, or countless other things. This latter is a transfer of the first person account to the third person account, and is a knowledge-transfer that can be explained through both the roles of empathy and, importantly, to having access to a shared value scheme (ibid.). Again, the public acceptance of this transfer of first person accounts points to the creation of a "common focus". In these views, we view the world together, construct our values, and work toward goals with unified action. In other words, "our collectivity concepts incorporate the concept of a plural subject" (Gilbert, 1992).

I concur with others: Artworks are testimony.

5.4.3 The Individual

But while we may be part of a "plural subject" we are still, in many ways, locked inside the confines of individual bodies—within our sense of "what it's like". Governed by David Marr's 2½ perspective wherein we see 3D but with the narrowed 2D information, our bodies experience the world thus funneled solipsistically toward us. This is the necessary condition of the I. The sufficient condition is my awareness of the sensory experience of the world as it is funneled to me e.g., my first person experience. Thus, because my sensory data is located proprietarily in my body, *and* I can be aware of that, *I therefore have an "I"*. I have a boundary between myself and the world; I have ownership of myself. And it is the "I" that makes art.

But, some people would say it is not only the "I"—a human—that makes art, for AI makes art, too. Or so some people say. I, though, would not, and my reason for that is as follows. AI makes images, but that is not the same thing as art, for it is essential that art be from the individualistic first person account, and not from a summative and generative machine.[30] Why? Let me formulate an argument for that claim.

Let's begin with the machine. Before LLM's, artificial intelligence was governed (during the 1950s and 1960s) by the programing of if-then statements (called "rule-based AI"), which obviously has its basis in symbolic logic. This though was revolutionized by neural nets in the 1980s, which encode patterns in weighed nodes, which function by registering the frequency of connections between nodes and thus the consequent patterns. This is modelled on the human brain, wherein the brain emphasizes and remembers on the basis of reiteration. LLM's are a slightly different version of inductive reasoning, because they are trained, whether it is supervised or unsupervised learning, to understand and give back information in language, predicting the next word in a sequence through probability functions.[31] The main point here is that artificial intelligence, in whatever version, works through inductive pattern recognition.

Now for the humans. The retort might be: Humans, too, work through inductive pattern recognition and also use neural nets, for brains are essentially the model for newer AI's such as LLM's. Thus, machines, like human brains, are essentially pattern recognition machines. The truth in that claim has been discussed in the first several chapters of the book wherein it was explained how our visual cortex uses neural stimulation for the retina in order to recognize things in the environment through the decoding of the patterns of those neural stimulations; the prefrontal cortex is also involved as are other parts of the brain such as the amygdala, basil ganglia, and the hippocampus, to name just some of them. The differences between us and AI, on this light, seem small even considering the difficulty AI's have in doing math.

True enough but not completely true. For the analogies don't spread across the comparisons evenly. So, now for the differences. *The differences are in the human body and the sensory experience of having David Marr's 2½ sketch and the affordances that are attached to that, which significantly come to me in gist.* The individual human's reality is simple: The world comes to me and it strikes me. It strikes me in the form of electrical impulses in light, or sonic pressure in sound, etc. I recoil or I lean forward; in other words, I have affordances—likes and aversions—that are based to some degree on my amygdala's fear-gauge function. Thus, things matter to me, I am "involved" (to quote Strawson). I want to draw close or I want to pull away.

This forces me, as an individual, to sort not according to how everyone else (or most everyone else) would sort, as would an AI constructed machine; but how *I* sort. While AI is going to decide based on the consensus view probabilistically

[30] This is consequent to the fact that art is not governed by mere depiction functions. See Sects. 4.2, 4.3, and 5.2.3.

[31] While I more or less had this history in my brain, I double checked it with ChatGPT(4), who (??) confirmed it. It was oddly satisfying to ask it to confirm its own origins of its cognitive processes. If only we could do that…

determined, I alone decide my affordances, and I alone react.[32] My bodily perspective is non-transferable, non-fungible, and never wholly given over to others. The absolute ownership of my perspectival experience makes all the difference, and it is what a machine could not have (not unless of course it had a limbic system…).[33]

This is the whole point of gist. We don't only see enormous amounts of data within the first 300 ms, *we bodily and emotionally respond to it*. Thus, cognition and emotion cannot be seen as separate functions. This makes the "involved" body an essential part of cognition. And since the making of art is a way of capturing our first person experience—capturing our "involved" body—art is a way the artist epistemically understands their place in the world. Thus, art is a kind of cognition.[34] And it is a function of singularly human cognition. To rephrase and pin the point more precisely: Art is the cognitive act of singular humans who, functioning as artists, rely on gist and affordances, and transduce first person experiences.

And now to champion the obvious: Without a body, there is no first person experience. AI's cannot use the part of the brain that activates bodily responses, the part that causes the body to move and respond to emotionally charged situations. That is not there. And so, neither is art.

But my opponent might counter with this: "those AI generated images look a lot like art to me!" Let me expand on this by explaining an orthogonal (but related) situation that describes the difference between our perception of natural facts in the world and manufactured objects. In the process of absorbing data we find constituent parts that are merely physical/concrete objects, and constituent parts that bear people's interpretations that are then incorporated into the objects. In other words, objects that are natural and objects that are not natural but instead human-made. For example, a little while ago I was looking at a carpet that had a floral pattern. That is a kind of perceiving that is distinct from looking at the sky, which I am doing now as I sit on the verandah. I am also looking at the railing of the verandah but that, too, is more like the carpet than like the sky. For when I view the railing I see not just metal, which could exist without human intervention or design, but I see instead metal that is constructed in a particular style: Simple, minimalist. It gives me the associations of a rational, clean, and functional world. And note: *Someone* has designed it so that I have those associations—someone gave me those quiet, almost invisible associations—those sentiments.[35] I am not, therefore, just seeing a physi-

[32] This is not to deny that we are influenced by others; that is an essential component of my decisions. But, nevertheless, the final decision is mine.

[33] This seems to make fear the root of all knowledge. I can't elaborate on that within the context of this book, but it does provide some basis for further research.

[34] I am setting aside the viewer's processing of art, which also involves first person functions, as the discussion under hand is the necessity of the artist as purveyor of the individual account in what counts as art.

[35] I use the words "sentiments" instead of "thoughts". To say that in the carpet or in the railing I am seeing the representation of someone's mental "thoughts" is perhaps not right. More correctly, perhaps, I am seeing the remnants of a sensibility; this is perhaps more correct to say. There was a minimalist, orderly sensibility in the railing and a relaxed, reassuring sensibility in the floral carpet. Designers are aware of these sensibilities that will conform to the expectations of the majority

cal object but the remnants of a mental one—someone else's mental one—as well i.e., the designer whose intention it was to "design a clean and streamlined railing". But when I look at the sky, on the other hand, I have no human scrim through which I view it, no scrim with a designed, additional and prior interpretation of reality that would constrain or corral my thoughts.[36]

Designed objects are distinct from artworks. There is a reason they are named differently and it worthwhile noting this. Designed items are closer to what AI's do, as designers must attend to the general standards of taste (read: majority). For example, if you ask ChatGPT what is "cold medicine", it will give you the over-the-counter categories including decongestants, antihistamines, etc. It won't give you my Aunt Pearl's remedy or Chinese medicine or anything else that hasn't garnered the approval of the vast majority of Americans. Likewise, if you ask a designer to design a table, you will probably get a table that conforms to most of our contemporary modernist standards; you won't get medieval, you won't get Chinese, and you won't get a completely individualized and unusual table no one has ever seen before. The majority rules, in both instances.

And though I might see the traces of human thought and intention in manufactured and designed items, artworks are significantly different. Because artworks are the transductions of the first person experiences of *singular* individuals[37] and incorporate not just sensibilities (designed in accordance with pre-existing standards for tastefulness) but also (and primarily) individualized attitudes, which are in accordance with the artist's own bodily experiences of the world, artworks become different from design. Artworks are singular. They present the point of view of an individual. Not of the aggregate group.

But while this difference between machines and humans may give us a normative rule e.g., I *ought* to care about the human individual's perspective, it does not give us a factual and pragmatic one. There is always a breach between the way things are and the way they should be, but contrary to the usual view it is the way things are that are often of particular value. This is the pragmatist rule e.g., we do *in fact* care about the human experience in art in ways that we do not with machine art. This is derivative not only of the definition of "art", but, relatedly, of its function as well. That function is largely pragmatic.

---

of the society. The designer could very well have uttered the sentence, "I need to design a clean and streamlined railing", but that would not necessarily entail much more than the desire to conform to usual standards of taste, which shouldn't be placed in the same category as a conscious commitment to a propositional statement. It would fall more into the category of an emotional desire to conform and to keep one's job.

[36] The usual ontological breakdown is between the material and the ideal—the object external to us and the idea "in our head" But that is incorrect. For the material is not always just material. Most material objects are, for most of us, objects that have already been sifted through human consciousness such that when we see them we are also simultaneously experiencing someone's thoughts. Pure objects e.g., nature, are relatively rare in our world.

[37] This is true despite the increased presence of group projects in the artworld, because those are formed on the basis of an agreed up point of view, which is derivative of an individual. I have a group project, talkPOPc, and it is formed like that.

As the recent decades of vision science and studies in psychology and neuroscience have shown us, emotion is an integral part of cognition. The first several chapters of this book underscored those findings. Thus, understanding is not merely naming the relevant facts, it is, rather, being able to put them into a cognitive framework that also has embedded in it a value system that can determine the relative weights of the different variables. In other words, a framework that can put one bit of data next to another bit of data and have them cohere. Why would the individualistic perspective of an artist be valuable in this enterprise?

We each, as individuals, often get things wrong. Locked in our sensorial bodies, we are also locked within the inductive practices of our minds. We frame things according to how we have framed them in the past. But then we shift and look at an artwork and suddenly the perspective of someone else comes into play and we see options we hadn't seen before. For example, I look at Reggie Burrows Hodges' "Electric Muthafukka" (2021) (Fig. 5.3):

Hodges is known to paint the canvas black before beginning a painting, and coupled with the fact that he is a black artist, some have been tempted to see metaphor, although he himself has never advocated that. Instead, his argument regards the formalist and emotive effect as the canvas sinks into its own depth. When I first look at this painting, I see not only a faceless black person sitting in a chair, but I experience that figure in the crowded, imploding density of the surrounding shapes as they encroach on the person from above and below. At first, that response is contraindicated by the somewhat firm placement of the legs that start to seem as though they are meeting the ground, but then on closer look I cannot find the feet any more than I could find the face; the feet seem to be sucked into the floor. The sturdiness of the figure now seems to be ingulfed much like the claustrophobic 1893 Édouard Vuillard painting of his mother and sister where the sister is being swallowed (more or less) by the wallpaper.[38] Hodges gives us a world like that, and I am glad I got a glimpse of it. After looking at it, my body felt less serene, less sure of itself, more cognizant of the world at my doorstep—the air next to me seemed weightier, the objects brightly colored, but not necessarily on my side… I had been changed.

Each artist's work is singularly quirky: No two artists do exactly the same kind of work. This is worth reflecting on for a moment. How does it happen that "the world" e.g., that univocally defined and monolithically existing entity is, in fact, so singularly experienced by each of us? The David Marr 2½ sketch goes far in explaining that but we need to go further. The answer is in induction.

There is an utter distinctiveness that each artist brings to the table of social intercourse—in fact, what all of us, artists or not, bring to the table is utterly distinct. No two people look exactly alike; people have distinct gaits, tones of voices, etc. Each life births distinctive inductive patterns that are valuable not just to the owner of those patterns, but heuristically to others. Why? Induction and pattern recognition comes into play here, too, in two ways. The artist has had those life experiences that are made sense of inductively as remembered patterns, and then handed over those

[38] See https://www.arthistoryproject.com/artists/edouard-vuillard/mother-and-sister-of-the-artist/

Fig. 5.3 Reggie Burrows Hodges. *Electric Muthafukka*, 2021. Acrylic on canvas. 76¼ × 65 in.; 193.7 × 165.1 cm. 76¾ × 65½ × 2½ in.; 194.9 × 166.4 × 6.4 cm (framed). (© Reggie Burrows Hodges. Courtesy the artist and Karma)

inductive, lived experiences to others in the best way that those inductive experiences can, in fact, be handed over to others i.e., in an artwork. It opens a window into another's perspective and thus presents options not previously available. We see in art another's choices, another's attitude, another's way of framing the problems of existence. I saw Hodges' work and windows opened for me. I felt more constrained, but oddly, too, more free. I wasn't sitting in a mythical quasi-Newtonian empty space, but instead head-butted up against other competing entities, friend or foe I do not know. So, I recognize now the co-existing world, and respond.

It is in those kind of experiences that we see potential solutions. Different ways of framing the sandstorm of data as it comes. And, often, other potential emotional choices. Thus, art gives us not images but attitudes with which to view the world.[39]

AI can't do that.

[39] Again, this does not make the artwork isomorphic with the artist's intention. While the artist's intention is necessary to take into consideration as one factor among others when understanding the meaning of the artwork, it is not the sole determinant. See Sects. 2.3 and 5.3 for the distinction between conscious and nonconscious intentions as well the publicly constituted meaning. As a further example, an artist might well *consciously* intend to make calm and serene art, but as this is not in fact her attitude to the world/her first person experience of that world, it would be the attitude that would triumph by demonstrating itself through such things as brushstroke, texture, color

5.4.4 Humane Knowledge

My understanding of the world is necessarily a parsed understanding. This is because I have parsed the incoming data—edited it—according to my inductive, habitual practices. In this, I am a *very* partial understand-er. And yet, hilariously so, I am so constituted as to think my understanding is quite adequate, in fact, quite correct! This seems to be what Hobbes was ridiculing when he was discussing the universal beliefs of mankind in regard to their own opinions: "For such is the nature of men, that howsoever they may acknowledge many others to be more witty, or more eloquent, or more learned; Yet they will hardly believe there be many so wise as themselves: For they see their own wit at hand, and other mens at a distance" (Hobbes, 1881). We see our own opinions "at hand" as a consequence of the fact that we each believe, quite naturally, that our perception is correct, for "seeing is believing"; others' perspectives seem quite foreign and riddled with error in ways our own do not. Thus, we each seem wise to ourselves. Hobbes' views could have been underwritten by Marr's, for Hobbes' empiricism carried with it individualized epistemological relativism, which is consonant with Marr's 2½ perspective; the world comes to each of us solipsistically. But for Hobbes, it was also the case that the world funneled to each of us in ways that resulted in not only solipsistic certitude, but with its inevitably consequential conflict.

Art is a way out of that solipsism. It is a way to see not only my own perspective "at hand" but others' as well, and it is so quite naturally. This is because art is an embodied form of perceptual conversation—a way of conversing with one another that is embodied and not linguistic. Our bodies have evolved to know how to read low-level features instantaneously and are thus receptive to the non-linguistic attitudes and views of others in general. We are therefore also able to do the same things when we interact with artworks. And thus art is evolutionarily effective as a form of conversation.

Although art traffics in the same milieu of bodily communication as does the rest of our lives, it has an added benefit because it comes to us outside the bounds of everyday reality. The latter is the world where navigate each other (often with avoidance or conflict), move information around, or proprietarily hand things back and forth between each other.[40] *That* world is the world of named objects. Art is an extremely effective form of conversation because it has left that world aside. For language has its limits; philosophers have usually expounded upon the glories of our language capabilities, but its limitations also deserve to be noted. However smoothly and easily acquired this kind of semantic knowledge seems to be, it is not a full account of what it is that we know.

---

choices, etc. Thus, some things that were meant to get said do get said, and other things that were not meant to get said do, nonetheless, get said anyway.

[40] See also (Buchak, 2021) for a discussion of how we navigate others' views in ordinary, non-aesthetic conditions.

In other words, not all knowing is governed by naming, and the class of things we know about the world is not identical to the class of things that we can name; we know much more than we can name.

This differs from usual dichotomies. For example, psychologists want to say that information is brought into our bodies in two ways: consciously and unconsciously. Though I have made use of this dichotomy throughout the book, I would like to now amend it by being more precise. I would instead divide the field into (1) data recognized by the semantic memory, and (2) data experienced and yet not recognized by the semantic memory. The former is the category of previously concatenated objects, which are hence part of the third-person nameable world; this would include the telephone next to me, the desk, the wall, the pictures on the wall and all the objects that I can name. Our mind recognizes patterns, and semantic naming is a culmination of that pattern recognition process.

But semantic naming is also a way out of that process of pattern recognition. We name things in order to stop the flow of patterns that come to us as a plethora of data. With naming, we circumscribe the data in question relative to a particular, *useful* object, pull it away from both other data (other objects which are not for that moment the focus of one's attention) and the constant flow of incoming data, and take it, as it were, out of both the constant flow of information and out of time. We name it. It is now static. It's a fungible quality that can be manipulated, exchanged, pointed to, discussed, and ultimately be converted to a quantity of some measurable kind.

But let us look for a moment at those things that are not named, because that is the really interesting category. For those things that are experienced and yet not recognized by the semantic memory are things that are often nonetheless important. For example, a member of this category would be "the flow of the familiar". We all know the feeling of familiarity of a place—the way the hallway feels as I walk down it, wherein my recognition is not of this object or that object but just the overall "feel" of the situation. This is the basis of our attachment to many objects—my favorite chair is intimately familiar to me but the qualities that make it so are only partially accessible to my conscious brain; it is only partially accessible to my ability to circumscribe those qualities and name them. That's what people are referring to when they resort to "I don't know what it is exactly about it that is familiar I just recognize the *feel* of it." This is not quite the same as the psychologist's "unconscious" category. Rather, it is a category of things experienced but not named. In general these are the low-level features of the environment, and I suspect that when I re-experience those familiar settings, the present data connects with those memories stored as episodic memory, which stores such things as what I had for dinner last night, or what I did on my vacation, but I think it must also connect with those parts of my brain that process process itself e.g., procedural memory (the basil ganglia).

This category of unnamed but experienced entities is a large part of the furniture of the mind—the "what it's like" of *my* world; the world as I experience it. It is as though the realm of the generalized world of knowledge comes into our bodies more democratically, without being singled out, ontologically circumscribed, and named.[41] It allows me to know "the feel" of walking down the hallway, and that kind of knowledge is important. In fact, many of the most important things in our lives are in that category: I know "the feel" of my mother's presence, and yet I cannot name the attributes that constitute that feel.

It is this that art gives us: This is knowing, not naming. Certain things only exist at certain speeds. In the same way that traveling near the speed of light shifts time, allowing the astronaut to see into the future as he or she returns to earth, so experiencing the world around us at a different than normal speed can allow us to perceive a reality that is not existent at normal ranges.[42] Looking at art can be one of those time shifts. Slowed down to almost a dead stop, our perceptual faculties are peering so closely at minute details that almost everything gets seen and taken in. It is similar to a meditative state where the mind, in its willed passivity, seems to experience the very atoms of life.

Art can, similarly, teach the observer to experience deeply each trace of human intention that the art encapsulates. While language and naming gives us a culture of pre-selected choices that serve as an editing device that grants speed, the knowing we get from art is much broader. Because art so fully relies on gist perception—both the original gist perception of the artist's as well as the gist perception of the viewer—art is a way of practicing perceptual gist. And what would we gain by practicing gist perception through an artwork?

We practice non-solipsistic perception. Let us briefly recap the facts. When we step in front of an artwork the first several hundred milliseconds give us the whole scene object, which of course includes the low-level features of illumination, texture, color, etc. It also gives us immediate affordances of desire/aversion. Thus, art shows us these things that, properly selected out and emphasized, give a point of view—a gist of the scene. They are the artist's point of view and now that has been handed over to the (observant) viewer. It is a way of giving a different kind of reality; a re-shaped reality. And the artist gets there by importantly *not* seeing something only in terms of the names that have been attached to it. It is seeing it apart from those names. The names will only get in the way of the experience.

This lack of role for names begins at the beginning, as names do not play the primary role in the fine-grained experience of making art, which then is repeated when the perceiver experiences in the artwork the reiterated gist experience of the artist; this effect (e.g., that limits a role for names) is compounded with the perceiver's own low-level perception and, importantly, their affordances that are embedded in the gist experience. The latter are, of course, the strong pulls and strong aversions. This shouldn't surprise: It is a well-trodden fact that art stirs emotional responses. It is also the fundamental mechanism in artworks, and the reason

[41] Obviously, what I am describing are low-level features; I recognize "the feel" of the hallway because I recognize the low-resolution shapes, the illumination, etc.

[42] To use an example from MIT Technology Review: If one twin is an astronaut, upon his return from space he will have aged 8 years while his twin that remained on earth has aged 50 years. Thus, upon returning, he is seeing into his future as he looks at his twin. https://www.technologyreview.com/2019/12/07/65014/how-does-time-dilation-affect-aging-during-high-speed-space-travel/

art is an embodied conversation. Because gist has embedded in it the mechanism of emotional affordances, and instantaneous reactions, the communicative function of artworks are not an additive feature but built into the very skeletal structure of the enterprise.

Thus, art steps outside those bounds of named entities, and we are able to experience perceptually freed low-level features with all of their affordances and newly presented attitude in place. In art, we are liberated to experience that reality outside reality, so to speak, for reasons I stated earlier in Sect. 5.3.3: The point of view presented by the artwork is experienced by the viewer at a remove from real-life consequences. We can fully and openly experience someone else's first person experiences, their point of view, in relative safety. In the example provided earlier, I can experience "without having to get too terribly close to the consequences I might suffer if I were roommates or lovers with him. It is a way of trying-on new attitudes on a trial basis, so to speak. Safe, novel, and a new way of approaching the world.

This aspect that allows the viewer to be able to be open to these new sorts of affordances and ways of framing the world is worth a closer look. Because the lack of real-life consequences allows for a psychological (although provisional) vulnerability to those new options, this is itself is enough to create a sense of trust in the point of view. For example, I have been surprised when watching Wagner's "Ring Cycle" that I had become, in the process of watching and hearing it, completely accepting of the sexual relationship between Siegmund and his sister Sieglinde.[43] In real life, where consequences are painful, such an option would not prevail. But Wagner's musical presentation of the quasi-Nordic tale of the gods is so utterly and viscerally moving, that intellectual opposition gets discarded and replaced by a yielding when we join him in that "collective belief".[44] Of course, the end of the cycle (i.e., the fourth opera in the cycle), pain does descend as a consequence of their relationship (perhaps…views differ on this), and there is the ultimate downfall of the gods. The "collective belief" has shifted and we have joined him there, too. This is why art is an attitude one holds toward things. It is knowledge, not naming. And that is crucial, for it is the bridge to belief.

Visual perception in general is extremely important to action because perception sets in motion the processes that lead to belief, which in turn leads to decision-making and thus action. In what way is the experience of Wagner or of Degas or of the contemporary artists Hodges or Brown, a kind of belief, and hence a claim to knowledge acquisition and what do we get out of that kind of experience? We get a new attitude. It is not facts in the world that have been given to us anew. The difference is in what one notices, how one approaches the world, how one frames the world—with what attitude, with what sense of possibilities, with what set of values. Obviously, all those things can lead to action. Gaining new options for framing the world is a kind of learning becomes it opens new possibilities for framing problems and hence devising solutions.

[43] The same effect holds true for Game of Throne characters Cersei and her twin brother Jaime.r

[44] Well, yes, one can almost hear the echoes of Plato's objections to artists.

And it is pleasurable. Why? The answer lies in the function of art, and the pleasure that we derive from that function. Aristotle, in the first sentence of the *Metaphysics*, writes that we get pleasure from learning and figuring something out. But the greatest pleasure doesn't necessarily come from the most abstract learning. Art does not lead us into the realm of abstract generalities, but rather maintains us in the realm of particular concrete experience—the sensorial data—and gives us an understanding of ourselves as embodied living beings and as others as similarly so construed. Being open to allowing others to pick and choose what parts of reality to notice, to name, and to value, is to be given a new reality. We are then open to the valuation of new perspectives.

As we are all partial knowers, art is the conversation between us that makes a new sub-set i.e., someone else's subset, of the data both legible and meaningful. As recipients of others' creations, e.g. as viewers of art, we see the world through the artists' transductions—through their scrim—and hence are able to differently parse reality and understand reality itself more fully than before. We are able to parse it differently because we've been exposed to the very direct epistemological choices conveyed by someone's art—to someone else's choices. Art matters because it gives us the pleasure of knowing ourselves, others, and reality itself.

The pleasure component is important. And note: The pleasure stems from the learning. The delight that we get when we have been exposed to someone else's choices, someone else's framing and attitude, gives options to us and gives potential solutions to our, otherwise, solipsistic and thus impoverished views. We have learned new attitudes to bring to the table, new ways to slice the pie.

Much has been written about art being pleasurable, but it has always been construed wrongly. It is not pleasurable because of the sensorial property of art. It is not the same thing as a warm bath. It is pleasurable because it is a kind of learning, and *that* is pleasurable. We are so constituted as to want to learn. And art is a kind of learning.

With that kind of view of cognition, it is obvious that art is training the mind to think. For art not only selects things out for us and forms those things as entities worth recognizing, it also tells how we should be responding to those things: With delicacy, aggression, approval, disgust, fear, etc. It tells us what to focus on and how to value it. It is full-brain comprehension; by relying on reiterated gist perception, activating the responses to affordances provided in low-level features, and encouraging an openness to others' first person experiences, art gives us a non-solipsistic way to parse the world, to learn, to feel pleasure. Thus: We are sieves. Porous to one another, influencing constantly one another's thoughts, attitudes, beliefs, judgements, and self-evaluations. We are, in a very real sense, constituent of one another.

Thus, art is our way of talking to one another and telling each other about other options, other ways of looking at life, other ways of responding and thus ways of solving problems. And it is able to do this because art is practicing perception in a social arena; it allows us to literally practice gist perception. Because it relies on low-level features, it is a way of us telling each other what to notice and how to value it (what attitude to hold toward it) in ways that are not linguistic, and hence, potentially antagonistic.[45] And with sympathy for another's view, we enlarge our own.

Art is therefore a social practice. It is the universal method that humans have used for tens of thousands of years that allows us not only to articulate our own individuality (if we are artists) but perhaps more importantly, as perceivers of art, to experience art and enlarge our gist experiences through the exposure to artists' gist perceptions and, thus, their associated affordances and attitudes toward the world. Knowing that affordances can be altered, even slightly, is a window onto other possible viewpoints and hence other possible solutions to life. Art is conversation, and it is our human—humane—way we have of not only passing information back and forth but of opening up membership to a common viewpoint.

References

Adolphs, R., Tranel, D., & Buchanan, T. W. (2005). Amygdala damage impairs emotional memory for gist but not details of complex stimuli. *Nature Neuroscience, 8*(4), 512–518.

Antony, L. M. (1987). *The Varieties of Reference*. Philosophical Review, *96*(2), 275. https://doi.org/10.2307/2185160.

Antony, L. (2020). Feminism without metaphysics or a deflationary account of gender. *Erkenntnis, 85*, 529–549.

Basu, R. (2019). To avoid moral failure, don't see people as Sherlock does. https://philpapers.org/rec/BASTAM-3.

Bermúdez, J. L. (2007). *Thinking without words*. Oxford University Press.

Buchak, L. (2021). A faithful response to disagreement. *Philosophical Review, 130*(2), 191–226.

Carpenter, M. (2010). Social cognition and social motivations in infancy. *The Wiley-Blackwell Handbook of Childhood Cognitive Development*, 106–128.

Cath, Y. (2019). Knowing what it is like and testimony. *Australasian Journal of Philosophy, 97*(1), 105–120.

Evans, G. (1982). *The varieties of reference*. Oxford University Press.

Evans, G. (1991). *The varieties of reference*. Oxford University Press.

Fuller, S. (2002). *Social epistemology*. Indiana University Press.

Goldman, A. I. (1999). *Knowledge in a social world*. Oxford University Press.

Goodman, N. (1976). Languages of art: *An approach to a theory of symbols*. Hackett publishing.

Gibson, J. J. (1977). The theory of affordances. *Hilldale, USA, 1*(2), 67–82.

Gilbert, M. (1992). *On social facts*. Princeton University Press.

Gilbert, M. (2007). Mutual recognition, common knowledge, and joint attention. *Hommage À Wlodek: Philosophical Papers Dedicated to Wlodek Rabinowicz*, 1–21.

Hobbes, T. (1881). *Leviathan, or, The matter, forme, & power of a common-wealth ecclesiasticall and civill*. Oxford: J. Thornton.

Jones, S. (2019). Testimony through culture: towards a theoretical framework. *Rethinking History, 23*(3), 257–278.

Lackey, J., & Sosa, E. (2006). *The epistemology of testimony*. Oxford University Press.

Legerstee, M., Markova, G., & Fisher, T. (2007). The role of maternal affect attunement in dyadic and triadic communication. *Infant Behavior and Development, 30*(2), 296–306.

[45] This is perhaps the solution to Hobbes' worry regarding our inability to see the world from others' perspectives, which, for him, led inevitably to war. Perhaps art is the way out, not authoritarian government.

Lewis, D. (2008). *Convention: A philosophical study*. John Wiley & Sons.
Malfatti, F. I. (2021). On understanding and testimony. *Erkenntnis, 86*(6), 1345–1365.
Mills, C. (2007). White ignorance. In *Race and Epistemologies of Ignorance*, ed. Shannon Sullivan & Nancy Tuana, 11–38.
Moss, S. (2018). *Probabilistic knowledge*. Oxford University Press.
Nagel, T. (1980). What is it like to be a bat? In *The language and thought series*, 159–168. Harvard University Press.
Peirce, C. S. (1986). The fixation of belief. In *Writings of Charles S. Peirce: A chronological edition, 3*, 242–257.
Quine, W. V. O. (1980). *From a logical point of view: Nine logico-philosophical essays*. Harvard University Press.
Racine, T. P., & Carpendale, J. I. (2007). The role of shared practice in joint attention. *British Journal of Developmental Psychology, 25*(1), 3–25.
Rosen, M. (1989). Social Facts. Review of Social Facts, by Margaret Gilbert. *Mind, 99*(394), 308–310.
Shaw, R. E., & McIntyre, M. (1974). Algoristic foundations to cognitive psychology. *Cognition and the Symbolic Processes*, 1.
Shaw, R., & McIntyre, M. (2024). Algoristic Foundations to Cognitive Psychology. In *Cognition and the symbolic processes*, 305–362. Routledge.
Shottenkirk, D. (2009). *Nominalism and its aftermath: The philosophy of Nelson Goodman*. Springer.
Siposova, B., & Carpenter, M. (2019). A new look at joint attention and common knowledge. *Cognition, 189*, 260–274.
Tomasello, M. (2014). Joint attention as social cognition. *Joint attention Psychology Press*,103–130.

Index

D. Shottenkirk, *Art as Cognition: How Gist Reframes the Aesthetic Experience as Conversation*, Synthese Library 521,
https://doi.org/10.1007/978-3-032-00143-6

MIX
Papier aus verantwortungsvollen Quellen
Paper from responsible sources
FSC® C105338

If you have any concerns about our products,
you can contact us on
ProductSafety@springernature.com

In case Publisher is established outside the EU,
the EU authorized representative is:
Springer Nature Customer Service Center GmbH
Europaplatz 3, 69115 Heidelberg, Germany

Printed by Libri Plureos GmbH
in Hamburg, Germany